Sheep of the World

SHEEP
of the World

Kenneth Ponting

BLANDFORD PRESS

Poole   Dorset

First published in the U.K. 1980

Copyright © 1980 Blandford Press Ltd,
Link House, West Street,
Poole, Dorset, BH15 1LL

ISBN 0 7137 0941 3

Phototypeset by
Oliver Burridge and Co. Ltd

Printed in Great Britain by
Sackville Press, Billericay, Essex
Bound by
Robert Hartnoll Ltd,
Bodmin

CONTENTS

ACKNOWLEDGE-MENTS

Thanks are due to many people, not only those who helped with the present book but friends in the wool trade for forty years.

It is quite impossible to list them all, but among many organisations I would particularly wish to mention the International Wool Secretariat and the British Wool Marketing Board, and among individuals my secretary, Mrs Eileen Anderson, not only for typing the manuscript but, because of her interest in sheep breeding, for providing much help and criticism.

Photographs and Line Drawings

The colour plates are credited as follows:

Pete Addis and Jim Byrne 48, 49, 50, 52
Animal Photography 1, 8, 9, 10, 11, 12, 13, 14, 15, 16, 17, 18, 19, 20, 21, 22, 23, 24, 25, 26, 27, 28, 29, 30, 31, 32, 33, 34, 35, 36, 37, 38, 39, 40, 41, 42, 43, 44
Australian Information Service, London 3
Grace Firth 47
Crispin Goodall, 51
Merino Stud Breeders' Association of South Africa 4, 5, 6
New Zealand High Commission, 46
New Zealand Romney Sheep Breeders' Association 45

The black and white photographs on the undermentioned pages have been provided by:

Pete Addis and Jim Byrne 130, 138
Agricultural Society of Iceland 118
American Hampshire Sheep Association 49
Anglo-Chinese Educational Institute 96, 97

British Wool Board 53
Cheviot Sheep Society 48
Colorado Division of Wildlife 124
Columbia Breed Society 125
Continental Dorset Club, USA 49
Dalesbred Sheep Breeders' Association Ltd 56
International Wool Secretariat 101, 127
Interfoto MTI +Hungary 17, 115
Israeli Ministry of Agriculture 93
Douglas Low 61
Museum of English Rural Life, Reading 65, 133
Netherlands Ministry of Agriculture 121
Novosti Press Agency 85, 86, 89, 91, 135, 136, 141
Society for Anglo-Chinese Understanding 26
South African Wool Board 37
Swedish Embassy 119

The historic sheep were drawn by Anita Lawrence and the maps by Christine Taylor.

INTRODUCTION

Obviously it has not been possible to compress all I have wanted to write about sheep into a relatively short book, and it has necessarily involved leaving out a great deal including the adoption of what may appear to some a too personal approach. It is fifty years since, at the age of fifteen, my father took me to my first wool sale; for thirty-five years I bought and processed wool and, more recently, have become ever more interested in the animal that produces it. Because of this personal involvement I have occasionally allowed myself a small piece of wool expertise which I hope will be excused.

One major difficulty in writing this book has been the number of sheep breeds and the outstanding importance of two groups, the Merino and the British. In any study of the sheep the importance of these two groups must be stressed and I hope it has been done here. But the great variety must not be forgotten and space has meant that mention of these sheep breeds must necessarily be very short. There is also a second problem, I have, with a few minor exceptions, seen and examined all the varieties of Merino and British breeds listed and the notes on each are mainly based on this personal experience. With the third group, all other breeds, the position is very different and I have to a large extent depended for reference on a few outstanding volumes, especially Mason's *A Dictionary of Livestock Breeds* plus his *Sheep Breeds of the Mediterranean*, also Epstein's *Domestic Animals of China*, and the US Agricultural Research Service's *Sheep Breeding Team Visit to the USSR*.

Drawings of the leading existing wild breeds. **a** A Zackelschaf, still found in Hungary and Rumania. **b** Urial ram from the Ust-Urt region. **c** A long-tailed Eastern sheep. **d** Hair-sheep with Moufflon patch from Abyssinia. **e** A Moufflon ewe from Corsica. **f** A Moufflon ram from Cyprus. **g** Hair-sheep of ancient Egypt with goat characteristics. The best information on ancient sheep comes from drawings and carvings.

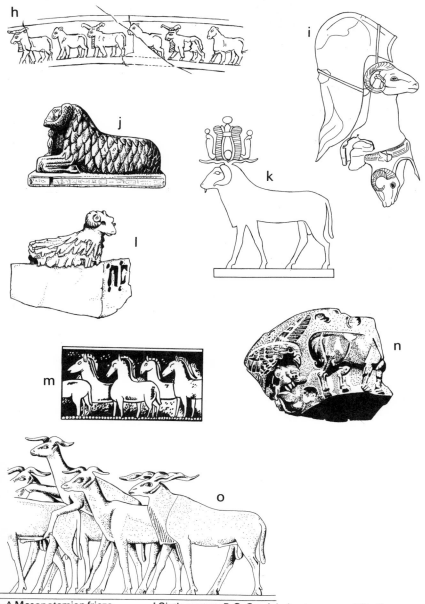

h A Mesopotamian frieze   **i** Sixth century B.C. Greek helmet.   **j** Similar type to the Zackelschaf, from Sudan.   **k** Egyptian sacred ram.   **l** Early Mesopotamian wool-sheep.   **m** Mycenaean sheep resembling the Cypriot Moufflon. **n** Stone bowl fragment showing an eagle attacking a fat-tailed sheep.   **o** Early Egyptian corkscrew-horned hair-sheep.

1 EARLY HISTORY OF SHEEP

The sheep was one of the first animals to be domesticated. Indeed, it could be argued that it followed the dog, which certainly came first and has, since palaeolithic times, had a unique relationship with man. Some would place the reindeer and the goat before the sheep but in world wide terms neither have had the same importance.

The domestication of both the goat and the sheep probably goes back to mesolithic (i.e. pre-settled agricultural) times. Remembering the unique hunting partnership of man and dog, man could then have changed from hunting the goat and the sheep to herding them. Maybe the special feeling man has had for the sheep dog arises from this time. If this was the case, the full domestication of the sheep would quickly follow as settled agriculture was established. Evidence suggests that this long and complicated series of events took place in western Asia and most of the breeds of today come from the sheep of that area although obviously wild breeds from elsewhere had an effect.

To go back to the wild sheep, the position is confused. There were some in the Pleistocene period when a kind of ancestral stock extended from western Europe to China. Today there are four main types of wild sheep: *the mouflon* of Europe, Asia Minor and western Persia; *the urial* of western Asia and Afghanistan; *the argali* of central Asia and *the bighorn* of northern Asia and northern America.

The mouflon today exists in two detached areas, one in western Europe, the other in western Asia. The former is restricted to Corsica and Sardinia. The complete absence of wild sheep from continental Europe is somewhat strange and has largely accounted for the generally held, but contested, view that all European domesticated sheep come from the urial. As far as the Asiatic mouflon is concerned it does appear to be agreed that they had little, if any, part in the foundation of the domesticated sheep.

Whatever effect the mouflon may have had in Europe there is no doubt that by far the largest number of domestic sheep today come from the urial which is currently found 'from Transcaspia through Turkistan, northern Persia and Afghanistan to western Tibet and the Punjab' (F. E. Zeuner *A History of Domesticated Animals*).

The argali, not to be confused with the arkal which is a race of the urial, is a large animal now found in a wide area ranging from Bokhara through the Altai Mountains and Tibet to northern China. It has certainly played a part in the domesticated sheep of India and the Far East but its importance in the overall position of the domestication of the sheep is very much second to the urial.

The bighorn found today in America, notably in the Rocky Mountains from Mexico to Alaska and also in northern Siberia, is the only wild sheep living which can be excluded from the ancestry of the present day domesticated breeds.

As will be made abundantly clear during the following pages, the sheep is more than a dual purpose animal—i.e. a producer of wool and meat. Early people would certainly have valued the milk, which only lost its importance with the domesticated and specialised development of the cow. If a pre-Neolithic domestication of the sheep is accepted, then in this early historical stage it would have been as a meat and milk producing animal that the sheep was valued, plus its skin when it was dead. The wool producing property, except in so far as they improved the warmth-containing properties of the skin, would not have come into the picture. There is no evidence that spinning was known before Neolithic times and most early Neolithic people wore skin rather than woven cloth. The realisation that the wool of the sheep could be spun and woven (also felted) into cloth played an important part in the wider domestication of the animal in Neolithic times but was not the cause of that earlier domestication as it seems that the early sheep may in many cases have been virtually woolless.

The most distinctive feature of the domesticated sheep as compared with the wild is its long tail, incidentally less noticeable in western Europe, Australasia and other areas where sheep are usually docked. Elsewhere sheep with long tails dangling almost to the ground were and are common, and they are often remarkably fat. (Herodotus noticed this peculiarity in sheep from Arabia.) At some point in domestication this change occurred and why long fat tails were prized has not been satisfactorily explained.

Domestication has also led to less need for horns and more and more sheep have become polled.

White wool became ever more in demand, particularly in Roman times, and the change from colour to white may be regarded as the most obvious result of domestication and proof, if any was needed, that from late Neolithic times onward the wool producing properties of the sheep have been paramount.

Origins of British Breeds

The next development can be exactly dated and comprises the great work of Bakewell and Ellman at the end of the eighteenth century. They were responsible for the selective breeding which created the great British Breeds and their work is fully described in Chapter 2.

Origins of the Merino

One major problem remains to be mentioned, namely the origin of the Merino sheep, the most distinctive of breeds. Its history in Spain after c. 1500 is well documented and is given in some detail later, but by that time it was to all intents and purposes the breed which in the late eighteenth century spread almost throughout the world. How did it come to be so distinctive a breed? The great difference between the Merino and other breeds and, to some extent, the immediate effect that the Merino sheep has on other breeds when crossed, has led many to think that there must have been some mutational change. But such an explanation really explains nothing and in any case does not give any indication where the mutation occurred.

Four possibilities suggest themselves. Firstly, the change could have occurred early in Asia c. 500 B.C., secondly it could have been the work of the Romans, thirdly it could have occurred in North Africa or finally in Spain, either in Moslem Spain c. A.D. 1000 or even later. To decide between these four possibilities in our present state of knowledge is impossible. Four distinguished scholars —three zoologists and one medieval historian—with whom I have discussed it, have all plumped for different answers. I am tempted to favour a post-Christian date, which rules out the first possibility and I doubt if Roman literature really supports the second. So this leaves one with a North African or Spanish origin. If one had to choose between them, a North African origin seems the most likely answer, with considerable development (or improvement) in Spain c. 1000.

To summarise, the six major developments in the history of the domesticated sheep are as follows:

1 Semi-domestication, when man helped by his dog changed the wild sheep into an animal that could be herded and moved with him as he hunted. **c. 10,000 B.C.**

2 Full domestication of the sheep in the main Neolithic period, at first mainly for its meat, milk and skin but soon the development of the fleece as spinning and weaving became widely adopted. **c. 5,000 B.C.**

3 Development of the white woolled sheep, mainly an achievement of the Roman period. **c. 100 B.C.**

4 The coming of the Merino sheep which may have predated the development of the white woolled sheep, but was most likely an Islamic (or Moslem) achievement. **c. 750 B.C.**

5 The work of Bakewell and Ellman in Britain which led to the so-called Long Wool and Down breeds which between them have dominated the sheep breeding picture as far as meat production is concerned. **A.D. c. 1800.**

6 Almost contemporary with the work of Bakewell and Ellman, the dispersion of the Merino sheep throughout the world. **A.D. c. 1820.**

2 THE MERINO

The Merino is the most important and most distinctive of all sheep breeds. Its early history has been briefly described; here we deal with the development in Spain and its dispersion throughout the world. The story of the development of the breed leading to the production of superfine wool remains today full of gaps. But first it is necessary to define what is meant by the Merino sheep. Sheep breeds of the world can be divided in several ways and, indeed, in place of the threefold division adopted in this volume one could make the division between the Merino and all the rest. This would be the case if one was looking at the sheep primarily as a wool producer, which was for many centuries its main object. The Merino is the breed that gives the fine wool, all the others in their many and varied forms, both pure and crossbred, give relatively coarse wool and are now primarily produced for meat although this was not always the case, and even today there are exceptions.

Spanish Merino
The original primitive sheep like many other animals had two coats, a fine inner and a coarse outer, and it still remains uncertain how the Merino sheep, where only the fine inner fibres are grown, was evolved. Here, however, it is reasonable to maintain that the ultimate development took place in Spain, but that is almost all we know. According to Ziegler in *The Merino*, 'the most plausible view is that the earlier Merino flocks recognisable as such were introduced by and named after the *Beni-Merines*, one of the North African tribes of the Berber

movement into Spain, as late as the twelfth century. It is quite certain that what became the Spanish Merino was unknown on the peninsular before that time, for the famous Moorish classic on agriculture in Spain (Abu Zacharia Ben Ahmed's *Book of Agriculture*) written just before the coming of the *Beni-Merines* makes no mention of any sheep resembling this breed.'

Fine wool from Spain began to appear in the European market towards the end of the sixteenth century. Previously there is nothing to indicate that the wool was better (in the wool sense, finer) than English wool. Ever since Merino wool has held the premier place, first coming from Spain, then for a relatively short period from Germany, and finally from Australia. The similarity of the Spanish and Australian countryside will strike all who know the two countries. Germany, both Saxony and Silesia, is different and here it was the specialised and careful breeding that led to the notable improvement. Generally, however, the Merino favours hot country—or perhaps one should say, is favoured there, as it is one of the few animals producing a valuable product that thrives in such conditions. (This need not be the case as I have bought fine Merino wool originating from places as far removed as the sides of Mt Cook in New Zealand, the Easter Island in the mid-Pacific and the highlands of Kenya.)

The wool follicles, or hair roots, of the Merino which may be called the fibre-producing part of the sheep, make an interesting study. (They can be compared with the machinery used in a modern fibre manufacturing factory.) There are in sheep, as in most other animals, two different types of follicles.

First, the so-called primary follicles always have a sweat gland opening into them, the secondary follicles do not. The secondary are more numerous than the primary, but in all sheep except the Merino in the ratio of about six to one. With the Merino the ratio is of the order of twenty-four to one. This means that the Merino sheep has about five times as many wool fibres on an equal skin area as compared to any other breed and this obviously accounts for the wool's fineness, although it still leaves uncertain the reason why it should have happened. There is something a little contradictory here. It would be reasonable to assume that the sheep that had to walk a long way in the hot sun, as would have been the case with the Spanish Merino, would have tended to have less fibres rather than more, but this in fact is not the case.

Migrations

When one studies the history of the Spanish Merino, the most notable feature is the so-called migratory system. The story of how these flocks moved each year from their summer pastures in the north to the plains of Estremadura and Andalusia, has often been told and has always fascinated the sheep and wool historian. Eileen Power in her well known volume *The Wool Trade in English Medieval History* wrote: 'Fine merino wool, of which the famous original came from North Africa, was established mainly during the Berber movement into Spain in the second half of the twelfth century. The Merino sheep formed the famous migratory flocks, the *transhumants*, which Don Quixote was to meet centuries later on their great annual trek from their summer pastures in the north uplands to the

Routes of the Merino Transhumantes in Spain.

winter pastures in the plains of Estremadura and Andalusia.'

The story of how these flocks moved each year has been best told in J. Klein *The Mesta* (the name given to the organising government department), but that volume written in 1920, two decades before Eileen Power's book, made no attempt to discuss whether it was this migratory system that led to the development of the fine Merino wool. There was, however, probably some connection. If the North African origin is correct, then the sheep had to walk partly for food and partly because their owners moved from one site to another. This nomadic life style could have led to the wool becoming finer. Then, when the animal was transferred to Spain, it may have been that unless this system was maintained, the wool lost its distinctive character. It would follow, therefore, that in a more developed civilisation the

walking, at one time somewhat haphazard, could develop into the standardised walks we find in Spain. We cannot know for certain, but it would be reasonable to suggest that continual walking could lead to mutational changes and the breeding out of the coarse wool and the development of the finer. It does seem that some such change must account for the development of the new fine wool, because there is nothing to suggest that it was done by specialised breeding.

The migrations fascinated observers and, to a large extent, annoyed the Spaniards. Don Quixote, saw them and thought they were an invading army. In this context it is perhaps worth noting that the distinguished, twentieth century writer André Gide, has an interesting passage in his *Journals* about meeting a migratory flock of sheep in the Mediterranean area. They were probably the Merino d'Arles which are moved from pastures near the Mediterranean up into the Maritime Alps each year. Many other breeds move regularly from winter to summer pastures, often as the snow clears from the higher land.

Klein, in his book, has given a detailed account of the way the Mesta controlled everything, quite surpassing in this any other of the many organisations that have attempted to bring order to the sheep and wool trade.

A major problem with the history of the Spanish Merino is that, as with the English medieval wool trade, we know so much more about the period of the trade's decline than its rise. By the time we have real evidence the Merino sheep had already become the animal that was to spread almost over the whole world.

The sheep travelled in detachments of ten thousand, guarded by fifty shepherds and as many dogs, with the mayorial or chief shepherd in front and these flocks possessed the right of pasturage over much of the kingdom. During these walks the sheep were said to keep admirable order, the assistant shepherds and the dogs keeping guard on the fringes, so preventing the stragglers from getting too far away or behind. As many as twenty miles were travelled each day and the farmers and the landowners through whose lands the sheep passed, suffered considerably and complained bitterly about the damage done to their property, but the right had existed from time immemorial and could not be stopped. However, by the seventeenth century the Mesta was fighting a losing battle against the ever-increasing opposition to its ancient rights, and as a result of this the fine woolled sheep of Spain began to decline in numbers. Nevertheless, a few of the famous old flocks re-mained and Spanish Merino wool certainly held its reputation in the European market. It was in fact during these years, perhaps partly because of the declining power of the Mesta, that the Merino sheep began to be sent abroad. During the early years of its ascendancy great care had been taken to see that no sheep should be exported and thereby enable other nations to produce fine wool. After the middle of the eighteenth century a few exceptions began to be made, the most important being in 1765, when the King of Spain presented three hundred Merino sheep to his relative, Prince Xavier, Elector of Saxony. These formed a stud which was carefully guarded and developed by selection and from it the Merino, assisted by further importations, spread over many areas. In Saxony there were no signs of a migratory system and improvements came entirely from selective breeding, which had been neglected in Spain.

At this point it is worth considering how fine the Spanish Merino wool was when it first became recognised in the late sixteenth century as the finest. A brief description of the ways in which the wool fineness, usually called wool quality, is measured, may be useful. In Britain, Australia and most parts of Europe, a system of numbers is used which relates to the size of worsted yarn that could be spun from that wool. The size of worsted yarn is measured in the number of hanks of 560 yards that weigh 1 lb. A 60s (24 micron)* quality wool is a wool that would produce sixty of these hanks (i.e. 60 × 560 yards) if

*One micron = $\frac{1}{1000}$ millimetre. Increasingly micron measurements are replacing the traditional quality numbers.

spun to the limit of fineness. Both mathematically and practically the system leaves much to be desired, but it has become well established and efforts made to introduce a more scientific method have not met with unqualified success. As it stands at present, all Merino wool is 60s (24 microns), or finer, all the coarser wools both from the pure bred sheep or from crosses between them, or even with the Merino itself, is known as Cheviot or Crossbred, which like so much else in wool nomenclature, is confusing.

The finest English wool of the sixteenth century came from the Ryeland breed and was about 58s (26 microns) quality. This wool was famous in Elizabethan times and known as 'Lemster ore'. We have pattern books from the seventeenth century which include actual wool samples and a considerable trade was done in so-called Spanish cloths but it is doubtful whether this always meant that Spanish wool only had been used. But from the samples that do remain and when identity can be reasonably assured, it would appear that Spanish wool then was between a 60s (24 micron) and 64s (21 microns) quality, in other words what is now known as a typical bulk type of Merino wool. It was only a little finer than the Ryeland but the trouble was that there was very little of the Ryeland type of wool available and most of the rest was 56s (28 microns) quality or lower. Spanish Merino, in the sixteenth to the eighteenth century remained the only bulk type of fine wool that was grown.

German Merino

Prior to 1765 there had been exports of sheep from Spain to her own colonies, some had been taken to South America by the conquistadores but these sheep were submerged and lost by mixing with the coarse woolled sheep that were also taken to the new world. Elsewhere, France had apparently obtained a few Merinos before 1721 by unknown means, presumably smuggling, and in 1723 the King of Sweden had obtained a flock. None of these exports led to any important development and the present to the Elector of Saxony of sheep from the Escurial flock, then considered the best of all, was the first deliberate and successful attempt to establish the breeding of the Merino outside Spain. It was one of the first successes of the planned dispersion of a useful animal on a world scale. Further exports followed from the same source in 1778 and other German princes vied with the Elector. Indeed, a great Merino-producing boom swept through Germany and in 1802 there were said to be four million sheep on the far side of the Rhine.

The German breeders concentrated on improving the quality and they spared no expense and certainly succeeded in establishing the finest sheep studs in the world. When the Merino sheep was taken to its present home in Australia, these German stud farms played a vital part in establishing the main flocks. Without them the great task of building the Australian sheep industry of the first half of the nineteenth century could not have been accomplished.

There are now comparatively few sheep in what was once the great sheep farming area of Saxony and Silesia, mainly in East Germany and Poland, but remnants can be found and will be detailed in a later section of this book. However, the fact of this disappearance

Merino breeding rams in Western Hungary.

1 Head of Australian Merino.

2 Shorn Merinos in pens in Australian landscape.

3. Australian Merino Sheep.

perhaps makes it desirable that something further should be said of the vital part that Germany played in the ultimate growth of the Merino sheep. For a time in the early decades of the nineteenth century, Saxony and, a little later, Silesia, produced the finest wool in the world. Later, the remnants of these flocks were found in Hungary, and firms needing this particular variety of Merino wool, bought their requirements from there well into the present century, and I remember seeing these wools in the 30s and late 40s.

These German breeders changed Merino wool from what would have been a 60/64s (22 microns) quality into the really superfine 70/80s (18 microns) of the great Merino flocks of the nineteenth and twentieth centuries. The difference being in the nature of 22 microns reduced to 18 microns. In practice much of the wool produced by the German growers was almost too fine and consequently at times a little lacking in strength, but this does not lessen the importance of the work that was done. These fine flocks, of which a few still remain in Australia, for example at Valleyfield in Tasmania, are of vital importance for stud purposes. When a commercial grower wants to get finer wool he buys a ram from Valleyfield and is, in fact, renewing links with the German Merino.

The felting quality which incidentally is no longer as highly prized as once was the case, came as a result of breeding finer. The finer the wool the closer and smaller the scales and although scales are not the only cause of felting, they are vital because they prevent the fibres once they have migrated together, from moving apart.

The German breeders brought to their work a care and attention that had not been practised before. Considering the shortage of Merino wool at the time it is a little strange that they should have used all their talents to go entirely for fineness. An effort to increase the weight of wool produced would have seemed the more sensible thing to do. The reasons that now apply for having a few fine studs would not have been so much to the fore in Germany, as the breeding for fineness came well before the great demand for fine woolled sheep to take to Australia had arisen. In any case it is arguable that such sheep would have been better suited for their new environment if the wool had not been quite as fine. The German breeders were apparently going for fineness for other reasons than the purely commercial, perhaps because of a kind of competitive spirit between the mainly large and rich landowners who led the Merino movement. They certainly did not know that they would now be remembered chiefly because of the part they played in the setting up of the great Australian sheep farms.

There are many interesting accounts of the visits the Australian breeders paid to Germany to get their sheep, and the story of Mrs Forlonge is perhaps not as well known as it should be. In 1826 this Scottish lady decided it would benefit her elder son if he emigrated to Australia. Her husband at first appears to have regarded the project with disapproval but eventually agreed. Two sons were sent to Germany to learn about fine woolled sheep and their management. Mrs Forlonge accompanied them and stayed four years. Having learnt all they could they purchased some of the best German Merinos they could find and walked them across Germany to Hamburg. They then shipped them to Hull, walked them again, this time across the Pennines, and one son, William, sailed with them for Australia. Mrs Forlonge herself went later, with a further consignment of German Merinos.

English Merino

The considerable importance of the Merino flock of George III in the Australian development of the Merino breed has only recently become clear. Until the publication of H. B. Carter's *His Majesty's Spanish Flocks* in 1964 it had been thought that they played a rather small part in the founding of the great sheep farms of Australasia, but his book shows that what can only be described as Wellington's major removal—perhaps thieving would be a better word—did play a greater part than had previously been reckoned the case.

Two themes emerge from this movement of Merino sheep from Spain to England. First there is the attempt to utilise them in Britain and secondly, the effect of their transfer to Australia. In both cases one of the main objects was to obtain for England fine wool necessary for the manufacture of the superfine broadcloths, particularly those of the West of England, but also those beginning to be manufactured in the West Riding of Yorkshire. The problem of obtaining sufficient fine wool of this type had greatly worried both the manufacturers and the government, indeed they appear never to have understood why Britain had ceased to be the producer of the finest wool in the world.

Early in the seventeenth century the clothiers had persuaded the government, who at that time were sure that English wool was the best, to prohibit the export of this English wool—a grossly unfair action against the agricultural community. It was only towards the end of the seventeenth century that both the manufacturers and the government frankly admitted that Spanish wool had to be used for the finest woollen cloths. Consequently, there was a great demand for this Spanish wool and a considerable shortage. As a result fine woollen cloth in the eighteenth century was expensive, a yard costing about as much as an average handloom weaver could earn in a week. These handloom weavers would pass their own suits of superfine broadcloths to their heirs.

There were two main attempts to overcome the problem of rectifying the shortage of fine wool. The first was to encourage the growth of the Merino in England and the second was to export these sheep to Australia. The chief advocates of these different solutions were Sir Joseph Banks, who played a major part in the bringing of the Spanish sheep to England, thus supporting the first method, whereas Captain John MacArthur the pioneer of Australian sheep farming not unnaturally believed in the second. We know now who was right but what our opinion would have been at the time is more difficult to say. In 1828 when the House of Lords reported on the Wool trade, there were many people who did not realise that the future of fine wool growing lay in Australia.

The first destination of the Spanish sheep was George III's farm at Kew and we have good details of how they were spread abroad but we do lack any real

4   South African Merino.

5   Fleece of South African Merino showing the fine, dense fibres.

6 & 7   South African Merinos and lambs.

knowledge of the way in which they were used. We know that there were a number of breeders in England who were interested in them and Merino sheep for a time appeared in all sorts of places—on the chalk downs of the south, particularly in Sussex where it is said that they did have an effect upon the native Southdown sheep; in the more northern parts of England, where it is also said, but this is more difficult to accept, that they improved the Cheviot. There was certainly a very considerable contemporary pamphlet literature on the subject. Generally, however, it is difficult to see much sign of their having any effect on the resulting wool and it is quite clear today that no British breed has much, if anything, of the Merino about it. To anyone who has handled wool, the difference between the wool of the Down sheep of England, and they are the nearest approach there is quality-wise, and the Merino is considerable.

In Australia the position was, of course, different and many of the breeders who played important parts in the development of the Merino sheep, notably Captain MacArthur, did use Spanish Merinos from Kew. However, even there they were not as important as the German Merinos which, it is clear, were the main foundation of the Australian Merino flocks in the crucial period around 1830. The Spanish Merinos from Kew did represent the main method by which the native Spanish breed of Merinos unaffected by German breeding methods, reached Australia and both the founding fathers, MacArthur and Marsden, bought sheep there.

To return to England, there is nevertheless (as the details set out by H. B. Carter show—and they are supported by Trow Smith's comments in his *History of British Livestock*) the fact that for a short time there was a real belief that the Merino sheep could be used profitably. The enthusiasm however, does seem to have disappeared almost as quickly as it grew and the reasons for this have never been clearly established. Maybe the sheep did so badly that the absurdity of the idea soon revealed itself; maybe it became so obvious that Australia was the right place for Merino sheep that interest waned. If the former, the unsuitable climate (too cold) was perhaps the reason, but too much should not be made of this because the Merino sheep, although it prefers a warm climate, will do well under various different conditions. Whatever the reason, the English Merino sheep as a separate breed was not to be and what importance the flocks at Kew played was in having a part in the dispersion of the Merino to Australia.

French Merino

The coming of the Merino sheep to France calls for a short account. This branch of the Merino is known as the Rambouillet because it was at the famous Chateau there that the sheep were first taken, and there indeed they still remain, and a visit to Rambouillet is essential for any visitor to France who is interested in sheep breeding. It is usually stated that the Rambouillet Merino was a bigger animal than the Saxon Merino and that the wool produced was not as fine. As a result of this, when the import of Rambouillet rams is mentioned in Australian literature, it is usually inferred that they were used to keep the Merino wool rather coarser than it would otherwise have been, perhaps indeed, to attempt to improve its mutton producing properties. Today, however, a visit to Rambouillet would hardly confirm this view because the wool of these sheep is now very fine, what a woolman would call a 70s (16 microns) quality.

More will be said about the Rambouillet Merino later but in this general survey it should be emphasised that in many parts of the world, notably in much of Europe and most particularly in America, the Rambouillet Merino was widely used, especially in the nineteenth century and as will become clear when dealing with individual breeds, its influence on sheep breeding has been very important.

The Australian Merino

The genesis of Australia is well known. In 1784 the British Home Secretary, Lord Sydney, not knowing what to do with all the convicts that were so overcrowding the gaols, decided to ship some of them to Australia. When these convicts set out, and equally during the first ten years of the colony's unhappy existence, the idea that Australia was to be the great sheepland of the next century never occurred to anybody. The fact that Britain was desperately short of fine wool at the time and that the Spanish Merino flocks were now being dispersed, might have led to this idea if there had been a well informed farmer present. The first Merino sheep came from the Cape and had probably been brought mainly with the idea of providing meat for the almost starving inhabitants. However, by a remarkable stroke of fortune these sheep were seen by Captain MacArthur who was largely responsible for the encouragement of and the growth of the great sheep farms that were the main facet in Australian nineteenth century history.

MacArthur was the man essentially responsible for the first great advance of this new Merino wool trade, not only did he do vital work in America but also in the marketing arrangements that he made in England which, as already indicated, was at that time desperately looking for wool. The manufacturers naturally viewed the new supply with delight. In 1810, 83 bales were imported, in 1834, 16,279; in 1840, 47,025 and by 1850 no fewer than 137,177 bales came over. At last the English fine woollen manufacturer had an adequate supply and, which was even more important, a supply of wool that was longer in length than the Spanish or German Merino and consequently was better suited for the new combing machinery that was slowly being developed for the worsted section of the trade.

The great worsted industry of the second half of the nineteenth century could not have advanced as it did unless this new fine and relatively long wool had not been available. The arrival of the Merino wool in large quantities in the textile markets of the western world represented the major development in textile raw materials and in a way can almost be compared with the coming of cotton at the end of the eighteenth century and even perhaps with the coming of synthetic fibres in the twentieth. It was indeed during the years between 1850 and 1914 that the Australian wool dominated the market as never before and hardly since.

In attempting to understand how Australian Merino wool evolved to become the main source of world supply, two changes must be emphasised. Firstly, the

8   Leicester Longwool.

9   Lincoln.

10 Romney Marsh.

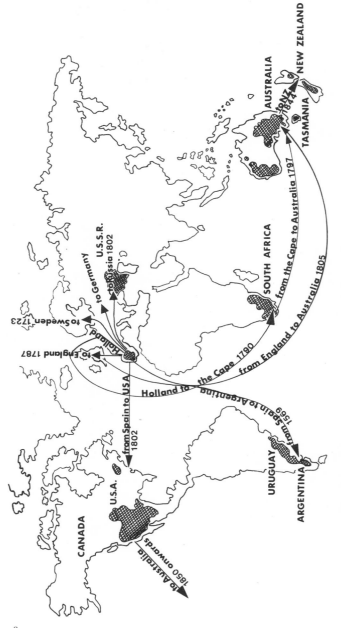

CANADA

U.S.A.

from Spain to USA
1802

to Australia
1850 onwards

URUGUAY

ARGENTINA

from Spain to Argentina
1569

Holland to Argentina

Holland to the Cape 1790

from England to Australia 1805

SOUTH AFRICA

from the Cape to Australia 1797

AUSTRALIA

to N.Z.
1844

NEW ZEALAND

TASMANIA

to England 1787

Holland

to Sweden 1723

to Germany

U.S.S.R.
to Russia 1802

KEY  Main centres of Merino breeding

wool became finer with the breeding of the Merino in Germany; secondly it became longer as a result of the breeding of the Merino in Australia. Such a combination is unique because if wool is grown longer it usually becomes coarser. From the point of view of the worsted section of the wool trade, which became more dominant, it was the increased length combined with maintained fineness that mattered. The development at the same time of machine combing made it practicable to manufacture all worsted cloths from fine wools.

Most of the early development work on the Merino sheep was done in New South Wales and it was there that the first stud farms were established which, combined with the new animals coming in, were to provide the necessary sheep to make Australia the greatest sheep country in the world. The second centre to come into prominence was Tasmania, then known as Van Dieman's land. Sheep farming began there in the early years of the nineteenth century and remained for some years of considerable importance, partly because at first, development was even faster than in New South Wales. During the years 1829 to 1831 more wool came to Britain from Tasmania than from New South Wales. This position did not last. Tasmania was used as a centre for the worst type of convicts and the amount of sheep stealing was quite fantastic. Nevertheless, sheep farming continued and Tasmania's importance in sheep breeding comes from the fact that the southern area of Victoria which, for more than a century has produced much of the finest Merino wool, was developed from sheep that had first of all been taken to Tasmania.

By the middle of the nineteenth century the lands around the coast of Australia had largely been taken over and it remained to move into the more central barren, semi-desert country. The fine woolled sheep based on the Saxon Merino, was not suitable for this area and a new type had to be developed.

The Peppin Merino

In 1858 on the edges of the great salt bush plain of the Western Riverina, the Peppin brothers began to put together a flock which during the next twenty years produced Australia's greatest individual contribution to sheep breeding. The Merino in Australia had until then been a small-framed, fine woolled sheep giving an average of 5-6 lbs of wool and not greatly changed except perhaps in the length of the wool, from the basic European Merino as derived from Spain and improved in Germany. It follows, therefore, that the Merino sheep before the Peppin brothers started their work, was about half its present size and, more important at the time, was quite unsuitable for the semi-drought conditions which prevailed in the central area into which the squatters were now moving because the richer lands near the sea had been occupied. One might have thought that the old Spanish Merino, habituated as it was to the dry plateaus of Spain, would have been a suitable sheep for this drier land but perhaps this particular characteristic of the Merino had been lost in earlier selective breeding.

Certainly the Peppin Merino represented a reversion from the superfine German, and among the many strains that the brothers introduced there must have been something of the pure Spanish. This new, medium woolled Merino sheep was clearly based on many

11 Blue Faced Leicester.

12 Wensleydale.

13 Border Leicester.

different breeds, not all of which have been identified. These strains were introduced as deliberate policy, and it should be remembered that the interior lands of various degrees of fertility were being occupied and that, as a result of this, many different varieties of Merino and indeed, a few non-Merino, sheep were arriving in Australia and were passing through the area where the Peppins had established their farm. Many of these could have been used, certainly French Rambouillet was brought in and although comparatively little has been said about it, some English long woolled blood, most likely Lincoln, was introduced. Whatever it was, this mingling of the races met with an extraordinary success as is shown by the fact that 85% of the present day leading studs in Australia are derived from the original Wanganella sheep and something like 90% of all the Merino sheep in Australia at the present time have Peppin blood in them. This success rested essentially on the sheep's capacity of withstanding the severe weather conditions of the salt bush, semi-desert country and yet preserving an increased weight of admittedly not so fine wool. It should be emphasised that longer wool was ideally suited for the bulk worsted trade, that was moving ahead so fast in Europe.

By about the 1870s the Australian sheep industry had been properly established. Its outlines were quite simple and although there have been many changes since, it is still possible after a further hundred years to recognise the founding fathers. Basically there is still the fine woolled sheep which is mainly serviced from the great stud farms such as two or three in Tasmania, notably Valleyfield, those in Western Victoria and those in the Yass country of New South Wales. They can all trace their history directly back to the Saxon Merino. At Wanganella one can still see the new Merino that the Peppins developed, but to some extent its importance over the past few decades has been replaced by the strong woolled South Australian Merino sheep which is, however, in many ways a similar type.

More important have been the changes that have arisen because the demand for fine woolled sheep began to decline in the last decades of the nineteenth century. Although there have been periods, especially after the two World Wars, when prices led to increased production, on the whole, particularly with the coming of the synthetic fibres in the mid-twentieth century, conditions have turned against fine wool growing. As a result in some areas there has been, when the land was good enough, a tendency to go either for cattle breeding, or to a somewhat lesser extent, to introduce British breeds and go for sheep that are mainly meat producing, and get coarse wool as a result. By the middle of the twentieth century British breeds, or crosses developed from them, account for the majority of the sheep in Tasmania and certain parts of Victoria. On the whole, however, New South Wales and more particularly, Queensland, has remained faithful to the Merino. Perhaps one should emphasise that in Queensland, although the wool is of good Merino quality, it is based more on the medium South Australian type which then goes finer owing to the climate, than on direct importations from the very fine stud farms already mentioned. West Australia has also been mainly Merino country and something is said about the particular strains developed there in the detailed accounts that follow. South Australia has tended to be the home of medium quality Merino wool and also of quite large quantities of coarser, non-Merino types. For the moment, but for how long in the future it is difficult to say, Australia remains the great home of that most fascinating of all sheep breeds, the Merino.

Main strains of the Merino today

Australian Merino
As described earlier in this chapter, originally from the European Merino, mainly German with some English—both of course originating from Spain.

Monte-Video (South American) Merino
This is the best of the South American Merinos. The wool produced is rather straighter and less soft in handle than the wools from Australia.

Mudgee (Australian) Merino
In the Mudgee district of New South Wales, distinctive, close, well grown type of Australian Merino, mainly deriving from the work of Cox, one of the better known of the early breeders. These sheep have played a leading part in the growth of the Merino in that area.

New Zealand Merino
The Merino sheep in New Zealand now represents only about 2% of the total clip but to anyone interested in sheep history it remains of considerable importance. There are few finer sheep areas than the Mackenzie Country, with the magnificent Mount Cook so completely dominating the countryside. The area also interests because the English author Samuel Butler farmed there in his young

days and made sufficient money in a few years to be able to devote his life to writing.

The story of wool growing in New Zealand differs considerably from that of Australia and can be divided into two main periods. At first it was essentially an adjunct of the great Merino sheep producing area of Australia and all the sheep there were of that breed. This period ended with the development of refrigeration but during these years the South Island was the main area and Australian sheep farmers, finding much of the best land in Australia gone, found in the Canterbury Plains a land eminently suitable for sheep farming. However, with the coming of refrigeration the Merino sheep largely disappeared, to be replaced mainly by English breeds, notably the Romney. Something about these British breeds in New Zealand will be said later.

Peppin (Australian) Merino
As stated in the introduction, this is arguably the most important of all the Australian Merinos. At Wanganella in New South Wales the Peppins developed a stronger Merino strain which proved ideal for the dry salt bush country further inland. The Australian Merino sheep trade could not have continued to expand after the better land was occupied without the development of this sheep. Nobody has ever quite decided what the Peppins did with the smaller, fine woolled Merino they started with, and some English blood must have been introduced, but from my own visit there I have thought from that Rambouillet blood was also introduced.

Queensland (Australian) Merino
Sheep came to Queensland after the

better land of New South Wales and Victoria had been occupied but the dry, comparatively barren lands proved ideal for the fine merino and this state has in the twentieth century been a major producer and less inclined to turn to other breeds than has happened further south with the richer wetter land. During my own days in the cloth industry I used considerable quantities of Queensland merino wools and they were notable for an all-round excellence. The sales at Brisbane were for a long time major centres for the sale of these wools.

South African Merino

A few Merinos seem to have been introduced to the Cape very early on but no real progress was made until Colonel Gordon imported Merinos from Spain of the Escurial stock in 1785. These appear to have been crossed with the hairy native sheep but after Colonel Gordon had died in 1797 his sheep were sold to a Captain Waterhouse, who took them to Australia. Success in real terms did not come until about 1803-6 and, more particularly, after 1806 when the British took control of the Cape. Colonel Caledon then began to introduce Merinos and, more important, in 1818 Lord Charles Somerset built up a pure-bred Merino stud at the Government Farm at Malmesbury. In this way the South African Merino sheep was developed and later around the middle of the century it spread into the dryer, semi-arid areas. Cape wool was first exported in 1820 but it was not till the middle of the century, after many experiments with different strains including the Spanish, the Saxon, the Rambouillet, plus some of the English breeds, that the definite South African

type of Merino was established. Although in fundamentals resembling the Australian, the wool from the Cape Merino has notable differences; it tends to felt quicker but will perhaps not stand up to prolonged felting as well as Australian. The wool taken as a whole tends at the present time to average a finer quality than Australian but there are probably no South African clips quite measuring up to the best from Australia.

South Australian Broad Woolled Merino

This strain of the Australian Merino was developed on the richer lands north of Adelaide and is widely used for crossing when the grower wishes to breed a larger animal with more weight and less fine wool. In Queensland particularly the dry land may mean the wool has grown too fine so that a certain 'tenderness' i.e. weakness of strength of fibre, may develop and South Australian Merinos are introduced to rectify this. In my wool buying days I would often see this happening with the larger, well known Queensland flocks. The Hawker family have been among the more important growers.

Tasmanian Fine Merino

One of the most important of the fine woolled Merino sheep which had their origins in those that Mrs Forlonge took to Australia from Saxony. It has played a great part in the growth of the sheep trade in Victoria and the few flocks that remain, notably at Valleyfield, are still vitally important in maintaining the fine wool strain in Australia. The wools from Valleyfield, Roc and, to a slightly lesser extent, Fordon delight the eye of the trained woolman.

Victorian (Australian) Merino
Is a very distinctive, fine soft handling
merino strain. The softness of handle is
very noticeable in the lambswool, which
often fetches the highest prices of all because it can be mixed with Cashmere to
produce the finest of knitwear. The Victorian fine woolled Merino was developed from the Tasmanian, but it now
has its own very distinctive qualities,
notably a particular brightness of colour
and as indicated above, an exceptional
handle.

West Australian (Swan River) Merino
The province was slow in developing. It
was not until towards the end of the
nineteenth century that this colony really
settled down and wool and sheep were
produced of a typical medium fine
Merino, giving good all-round wool but
with perhaps less distinction than the
other main Australian types.

West Australian (Murchison) Merino
The old gold-mining area of Western
Australia has produced a good medium
fine wool characterised by a remarkable
red colouring due to the soil, which does
however scour out to be a very good
white.

West Australian (Kimberley) Merino
Well to the north in West Australia the
dry, almost tropical climate has produced a fine woolled Merino, notably
soft handling and in some ways the most
distinctive of the West Australian wools.

Yass (New South Wales) Merino
One of the fine woolled Merino strains,
somewhat resembling the Mudgee.

Other Existing Strains

Apulian Merino
Seems to have originated as far back as

the fifteenth century but in its present
form is chiefly a late eighteenth century
development from Spanish Merino
crossed by local breeds with some Saxon
and Rambouillet blood added in the
nineteenth century. Mason in his *Sheep
Breeds of the Mediterranean* states: 'the
"noble" or improved breed of Apulia is
native to the tableland in the province of
Foggia, where the local breed was first
crossed with the Merino as long ago as
1435-42'. It is difficult to find any further
information regarding this early ex-
portation of the Merino.

Argentine (South American) Merino
Argentina was at one time a great producer of Merino wool but with the
coming of refrigeration the good land
became more profitably used for cattle
and Merino sheep are now of much less
importance. The originals seem to have
been based on improving the native
Creolli breed, first by Spanish and
Saxony Merino in the period 1815-75
and then, during the following fifteen
years, by the introduction of the Rambouillet.

Arles Merino (French Merinos d'Arles)
An interesting Merino, said to have been
derived from a local breed upgraded by
Merino. It was also improved late in
1921 by the Chatillonais variation of the
Précoce. This breed maintains the old
movement between summer and winter
pastures.

Black Merino (Portugal)
Is a fine wool breed used also for meat
and unusually for a Merino type, for
milk.

Danube Merino (South Romania)
Is another fine wool, dairy and meat

breed like the Portuguese but mainly white. It seems to have originated by a cross from the local Tsigai with the Merino, but is not now very important.

Delaine (American) Merino
This is the most successful of the American developments of Merino, not as well known, but certainly better than the Vermont.

Easter Island Merino
A flock of Merino sheep were taken to this isolated island many years ago and when I last had an opportunity of examining the wool, the tendency of the Merino sheep to revert to the curly, crimped type of the original Spanish-Saxon breed was very noticeable.

Italian Merino   *See* Apulian Merino.

Kenya Merino
During the twentieth century Merino sheep were taken from South Africa into the highland of Kenya and for a considerable time good and distinctive Merino wool was produced there.

Letelle Merino
This is a special type of Merino sheep with meat rather than wool the main object, introduced in South Africa from 1922-39 by J. P. van der Watt.

Merino-Landschaf (the German Improved Land)   *See* Würtemberg Merino.

Palas Merino (Constanta region, S. E. Romania)
A fine wool, and meat producing breed. The males usually horned, the females usually polled. Originated 1920-50 at the Palas Animal Breeding Station, Constanta, from Tsigai, Turcana and Stogosa, graded to Rambouillet 1926-34 and to German Mutton Merino 1928-42, with further Rambouillet and Stav-

ropol blood. The Romanian name is Merinos de Palas.

Polish Merino (Poland)
A fine wool, meat animal. Originally chiefly from Precoce in late nineteenth and early twentieth century. Named in 1946 and improved by the Caucasian Merino, 1952-57.

Portuguese Merino (South Portugal)
Fine wool, dairy products and meat. Originally from Bordaleira crossed with Spanish Merino (since fifteenth century), Rambouillet (since 1903) and predominantly Précoce (since 1929).

Précoce (France)
A fine woolled, meat breed. Males are horned or polled, and females polled. Originally from Spanish Merino imported 1799-1811, selected for early maturity. Herd Book breed created in 1929 by fusion of those of Champagne, Chatillonais, and Soissonais Merinos. The original of the German Mutton Merino, Polish Merino, Portuguese Merino. This is an interesting and important breed and, as the short detailed accounts of individual European breeds will show, has been very widely used for crossing.

Turkish Merino (North East Anatolia)
This was first introduced from France in the nineteenth century and was crossed with local Kivircik. Then in the twentieth century several European Merinos were brought in again—Hungarian, Württemberg and German Mutton—and there are now about 200,000.

Vojvodina Merino (North East Yugoslavia)
Wool is the primary product, then meat. The history of the Merino in this part of the world is interesting. When Croatia

South African Walrich Ewes.

was part of the Austrian Empire some Merinos were imported in 1771 (i.e. at the time of the great Merino dispersion), they seem to have been transferred to Bohemia or Hungary. Further importations took place in the nineteenth century and substantial flocks were established. They were decimated by the wars of the twentieth century and the break-up of the Austria-Hungarian Empire and by 1945 a rather poor, special type of Merino, the Vojvodina, survived. Recently efforts have been made to improve its size and fleece weight by introducing Precoce, Caucasian and Stavropol Merino blood.

Walrich (South African)
New Merino developed in South Africa.

Württemberg (South Germany, Alsace-Lorraine, France)
Produces medium wool and meat. Males polled or horned, females polled. Originally from Merino (imported late eighteenth and nineteenth century) by Württemberg Land; since 1950 the official German names are Merino Landschaf (W) and Deutsches veredeltes Landschaf (E), and the French is Race de l'Est à laine mérinos.

Important Derivations

American Rambouillet
Today the majority of the fine woolled sheep in America (although they must have partly descended from the original importation of Spanish Merino) have been greatly influenced by the Rambouillet, so much so that a distinctive type, the American Rambouillet, can be recognised. It is essentially a fine woolled and meat sheep and the main importations seem to have come from France between 1840 and 1860 with some additional German Rambouillet strains in between 1882 and 1920. A Breed Society was formed in 1889.

Comeback (Australian)

This is Merino crossed with a long woolled breed, usually Lincoln which had already been crossed with the Merino, hence the name 'Comeback'. The description is to all intents and purposes confined to Australasia and is more commonly applied to wool than to sheep. The Polwarth is probably the most common type of Comeback.

Corriedale (New Zealand)

Originated in New Zealand but is now widespread, and the Canadian Corriedale and the Soviet Corriedale should be and are here considered as separate breeds. The original Corriedale was based on a Lincoln × Merino, and the breed was developed between 1880 and 1910. Leicester sometimes replaced the Lincoln, but was probably not as satisfactory. The main idea of this cross as of so many others of the same type, was to get a Merino sheep that also produced useful meat. The most successful Merino/English Longwool cross that has been produced.

New Zealand Half-bred

Like the Corriedale, this New Zealand breed is a Longwool (Lincoln, Leicester or Romney) crossed with the Merino, but, at least in wool quality, differs considerably from the Corriedale. It is a little difficult to see why this should be. The phrase 'half-bred' is probably more a woolman's term than anything else and the distinctive qualities of this wool although fundamentally based upon the Merino-longwool cross, probably comes from the area where this breed is produced, namely the centre belt of the South Island of New Zealand. Here the half-bred is found often next door to those farms which still keep to the tradi-tional New Zealand Merino. The wool from this half-bred has a crispness of handle that at one time made it particularly sought after for making fine Scottish Cheviot tweeds. It is possible to see in the wool something of the distinctive qualities of the New Zealand Merino and it would therefore probably be best to describe it as a Longwool (Lincoln, Leicester or Romney) crossed with New Zealand Merino.

Polwarth (Australian)

One of the several attempts made, none except perhaps the Corriedale with complete success, to breed sheep that produced both wool and meat to everybody's satisfaction. With the Polwarth the Merino was crossed with an English Longwool type, usually the Lincoln.

Historical

American Merino

Many hundreds of Merino sheep, mainly Spanish, were taken to the United States between 1793 and 1811 and for a time the developing country was a major Merino producing part of the world. Rambouillet blood was also widely introduced. Several distinctive strains, even breeds, were developed, notably the Delaine and the Vermont. In the twentieth century the importance of the Merino greatly declined.

Camden Park (Australian) Merino

Important early strain of Australian Merino as developed by Captain Mac-Arthur, and including more English Merino than other early types but still basically Saxon ex Spain. A few Camden Park Merinos are still in New South Wales and are an invaluable museum-piece for all interested in this great breed.

English Merino

An account of the attempt to introduce the Merino sheep to England has been given in the introduction to this section.

Escurial Merino

Is an important former strain of the Spanish Merino. The Elector of Saxony's sheep came from Escurial.

Guadalupe Merino

Another former strain of Spanish Merino.

Hungarian Merino

Derived from importations made by Empress Maria Therese of 300 Spanish Merinos to the Imperial farm at Mereopail in Hungary. Towards the end of the period of importance for the European Merino, Hungary was probably the leading centre of production. It is not clear how far they affected the two Hungarian Merino sheep of today (Hungarian Combing Wool Merino and Hungarian Mutton Merino. During the twentieth century Hungarian Merino wool was still bought by West of England firms specialising in fine broadcloths, such as the coverings for billiard tables.

Infantado Merino

A former strain of the Spanish Merino.

Negretti Merino

Perhaps the most important strain of Spanish Merino. This is a convenient point to query how the Spanish Merino breed was divided and whether the various strains given here were really distinguishable. It would seem that the Spanish Merino during its most famous days, was a fairly autonomous group and that any attempt to sub-divide as is in fact attempted here, may well be un-

justified. The historical background of the Spanish Merino sheep has already been given.

Paular Merino

This was another important strain and it has been stated that the fleece of the Paular was the largest though not the best; the wool from it was reserved for the King of Spain's manufacturers.

Perales Merino

Former strain of Spanish Merino.

Rambouillet Merino

Originated as stated in importations made by Louis XVI from Spain to Rambouillet in 1786 where the direct descendants can still be seen. These sheep have played an immensely important part in the development of the Merino sheep throughout the modern world.

Russian Merino (Historical)

Merino sheep came to Russia in 1802 through the agency of a Frenchman, M. Rouvier, a bankrupt fleeing from his creditors who sought refuge in that country. He saw the steppes of Southern Russia and conceived a future there for the Spanish Merino sheep. He imported eighty rams into Sebastopol, obtained a land grant, and later brought in Saxon sheep as well. There seems to be very little historical information about what happened afterwards, how far these sheep have affected modern Russian Merino breeds. It is possible that the influence is in fact quite considerable, more perhaps than the present authorities in Russia are prepared to admit.

Saxony (German) Merino

It was in 1765 that the Elector of Saxony persuaded his cousin, the King of Spain, to present him with 100 Merino rams and 200 ewes from the finest flocks and this was the first deliberate and really successful attempt to establish the growing of wool outside Spain as a planned and official development. Further importations followed in 1778 when 100 rams and 200 ewes were obtained from the Royal Escurial flock. By 1802 there were 4,000,000 pure Merino sheep on the German side of the Rhine. As has been made clear German breeding methods quickly developed, between 1765 and 1820, a very distinctive, fine woolled sheep.

Silesian (German) Merino
Derived from imports made by Frederick II of Prussia in 1786 and later in the second half of the nineteenth century, more important than the Saxony Merino.

Spanish Merino
Generic term covering this most important of all sheep breeds. As indicated, there are a number of separate strains— *see* Escurial, Guadaloupe, Infantado, Negretti, Paular, Perales.

Vermont (American) Merino
The Vermont Merino was developed in America and was distinguished by its crimply skin, which it was hoped would lead to a bigger wool clip. Many were taken to Australia but it was found that the crinkly skin surface led to uneven wool and the sheep was not as suitable for the dry, near desert country and great damage was done by the Vermont to many famous flocks.

3 BRITISH BREEDS
History and Development
The history of sheep breeding in Britain is reasonably clear in general outline but more complicated in detail. First, therefore, the general outline. The first sheep probably came to Britain with those Neolithic newcomers who settled at various sites in the western part of the country, notably Windmill Hill in Wiltshire. The evidence of the bone structure of the sheep found at Windmill Hill suggests an animal of the Soay type and, indeed, in this remarkable breed we have a wonderful piece of animal archaeology. The next importation probably came with the Romans, who introduced a white faced sheep, signs of which can still be seen in the white faced, long woolled breeds of today, as also in the finer Cheviot and Dorset Horn. It is possible that two types of Roman sheep, one with rather coarse wool and one with finer, may have been introduced and these might have given rise to the two types indicated above, but satisfactory evidence to prove this point is lacking. Finally, about the middle of the sixth century the Vikings, when they settled in the Scottish islands and northern parts of the mainland, introduced a third sheep which today remains comparatively little changed in the Shetlands. This third arrival was considerably less important than the other two.

But sheep development does not only depend on breed; environment is at least as important. By the time we have more definite, although confusing, evidence we find in the Middle Ages (twelfth century) basically two types of sheep. A relatively long woolled sheep in Lincoln and other parts was a fairly homogenous type and as indicated, being mainly derived from the white faced Roman breed. Elsewhere, especially in the hill country, a rather varied group of sheep largely coming from the

Soay, inhabited the countryside, giving shorter wool and—this is important—finer wool. Poor country has always given finer wool than rich land and there is no reason to believe that this fundamental fact was different in the twelfth century than it is today.

There has perhaps been some confusion here, when dealing with the pre-eminence of English wool at the time. Two quite distinctive types of English wool were apparently sought, the one coming from the first type, long and relatively fine for its length but definitely not a fine wool. It should always be remembered that length with wool is relative to fineness. Thus today three inches is long for a Merino but very short for, say, a Border Leicester. This has, I believe, always been the case and English long wool was in demand for its length and for its relative fineness. The second demand for English wool, and I suspect this was the larger demand, was for the fine short wool which in a few areas—notably the marshes of Wales around Leominster—had been carefully bred from the native heath or forest breed which had itself derived from the original Soay type.

In the 1500s the agricultural environment began to change. Historians still argue about the extent of the enclosures but that some took place is not disputed. Even more important, the increase in the population meant an ever increasing demand for meat and all the pressures were to breed sheep more for meat and less for wool, with the inevitable result of larger animals and coarser wool. England then lost her position as the producer of the finest wool in the world to Spain but it should be emphasised that its reputation for long wool still remained.

The seventeenth and eighteenth centuries did not bring many changes. The export of wool was prohibited, such was the power of the cloth manufacturers of the time, but some wool was smuggled out, almost all of it doubtlessly long wool in which England still seems to have had something of a monopoly. Then in the period around 1800 the final change took place which led to the development of the breeds we know today. But one last importation should be mentioned. When Wellington's armies were fighting the French in Spain they, or their commanders, seem to have seized every possible occasion to send back to England quantities of Merino sheep and it was hoped that they would improve the native breeds and so prevent the importation of fine wool for superfine cloth. There was, as has been shown, never any hope that this would succeed. England in 1800 was about as unsuitable a country for the Merino as it is possible to imagine and the effect of the introduction of the Merino sheep into English breeds was minimal.

The Influence of Bakewell and Ellman

What was important at the time was the work of a few great animal breeders, Bakewell of Leicestershire and Ellman of Sussex. They took the sheep native to their areas, Bakewell the old Leicester, Ellman the small heath sheep of the downland of Sussex, and from them they produced the New Leicester and the Southdown. These two breeds have played a part in world sheep breeding second only to the development of the Merino. The New Leicester was crossed with other long woolled, white-faced sheep and, in particular in the Border country, gave rise to the Border Leicester —one of the great modern breeds. The Southdown was similarly used to improve the native heath sheep of the other down country, giving rise to several distinctive breeds, the Suffolk, Hampshire and Oxford Down for example.

Meanwhile other native sheep were improved by the new methods pioneered by Bakewell and Ellman. The Cheviot and the Dorset Horn were typical here; although some new blood was introduced, much of the improvement seems to have been the result of careful selection inside the breed. Further afield on the hills and mountains the effect of these new breeds took longer to make an impression—usually it was the Leicester, normally in its Border Leicester form that was used—but the mountain breeds, the Blackface, the Herdwick and the Swaledale, have remained nearer to the original than have the sheep of the richer and lower country. Further afield still, particularly in the northern islands, sheep like the Shetland are even less changed.

This great age of British sheep breeding had a world wide effect exceeding in importance that in the country itself. When the Merino sheep was exported in large quantities it went not only to Australia, which was to be its real late nineteenth and twentieth century home, but also to the USA and New Zealand where for a variety of reasons it did not settle permanently. In the USA population pressure, and in New Zealand the discovery of meat refrigeration plus a suitable climate, led to both countries largely giving up the Merino and turning towards the meat producing sheep, usually in the form of the fat lamb trade.

In order to do this they turned back to Britain for the basic stock which was either crossed with the Merino or, more often, bred pure. As a result this has meant that in New Zealand, for example, there are far more Romney sheep than there are in its own native corner of Britain. It follows, therefore, that the sheep breeds whose history is briefly traced in the following pages, have had a vital effect on world as well as on national breeding. Few animals have played such a part in agricultural history as the great British breeds such as the Border Leicester, the Southdown, the Romney and the Cheviot.

Longwool Breeds

Bluefaced Leicester

The breed is descended from the Border Leicester with some Wensleydale blood added to give it a distinctive quality. It was bred entirely for the purpose of breeding high quality crossbred ewes from hardy hill breeds and its increasing importance derives almost entirely from the quality of its progeny—the Greyface or Mule. This is one of the most important new breeds to be evolved in the late nineteenth and twentieth century.

Border Leicester

In the Border Leicester, the Border country has produced a breed worthy to rank with the Cheviot and the Blackface. This sheep is descended from the New or Dishley Leicester which was introduced into Northumberland in 1767 by the brothers, the Culleys. They did little crossing although a limited amount of Cheviot blood may have been introduced. Basically, they kept to the original Leicester but the sheep, perhaps because of the new habitat, gradually changed in type and a quite definite breed was produced and soon the annual sales were well attended. For many years the Culleys went back to Dishley and obtained their rams from Bakewell's own flock but after 1830 they ceased to do this, having evolved their own distinctive type.

The great importance of the breed does not lie in its wool or even its meat, rather in its quite outstanding importance as a stud ram. Its progeny mature early and fatten quickly, its crosses are excellent and it is no exaggeration to say that the Border Leicester sheep has come to dominate sheep breeding in Britain. There is little doubt that it will be increasingly used possibly more, however, in its Bluefaced Leicester version. In Britain the best known cross is that with the Cheviot ewe which produces the well known Scottish Half-bred. Farmers from the south of England go north either to buy their Border Leicester rams or even more frequently, to buy the Half-bred ewes to put with their own Down (now usually Suffolk) rams. The Border Leicester is also crossed with the Blackface to give the so-called Greyfaced, not as widely known outside Scotland as the Half-bred.

Few breeds of sheep are found over such a wide area. Travelling in Norway beyond the Arctic Circle I have seen a few, while not long ago the *Pastoral Review* of Melbourne published a photograph showing the sheep on Macquarie Island in Antarctic waters, with the caption 'The Border Leicester sheep are kept there for the purpose of providing meat for the expeditions on the island and reports indicate that they are thriving in this latitude. Blizzards lash the hills for about 340 days in the average year; these ewes were about to lamb on the 1st August last year, on which date a blizzard was blowing; they postponed their lambing for a month to take advantage of September's fine weather'. This last story seems a little unlikely but another photograph shows the craggy nature of the country as well as the misty weather which rules for most of the blizzard-free weeks of the year, plus four white-faced sheep facing the island climate with remarkable equanimity.

Between these two extremes, the sheep have done well in Australia and promise to do better in the future. In Australia the first Border Leicester stock was established in Geelong by L. C. Cochrane in 1881 and has proved its value as a crossing sheep. In New Zealand it is beginning to challenge the position of the Lincoln as a crossing sheep and, in its Half-bred form, even the Romney. Nearer to home the manner in which it has spread, again in its Half-bred form, to the downs of Southern England once the home of the old classic Down breeds stands out clearly to all who know the area.

British Oldenberg
A new development in Britain. The well known Oldenberg sheep was first imported in 1964, but has as yet made no great impression.

British Texel
Originally from the islands of that name off the coast of Holland, this sheep came to Britain in the mid twentieth century, but, like the Oldenberg, has not as yet made any great impression.

Cadzow Improved Ram
An attempt to introduce a new breed carried out in Scotland.

Cambridge

A new breed developed at Cambridge. Originally various strains were used, including the Cheviot, the Border Leicester, also the Finn ram, but it is now self-producing.

Colbred

A new breed developed during the 1950s in Britain, by Oscar Colburn from Border Leicester, Dorset Horn, Clun Forest and East Friesian.

Cotswold

Historically speaking the Cotswold sheep is one of the most interesting of all. The great historian Eileen Power, in *The Wool Trade in Medieval English History*, maintained that there were two main types of sheep in medieval England, one small producing short wool which was carded and used to make woollens, and the other larger, with long wool which was combed and used for worsteds. The smaller breed was kept on the poorer pastures of the hills, moors and dales and was found in the Welsh and Scottish border country, in the Yorkshire moors and the chalk downs of the south. The most famous of all the short woolled breeds was the Ryeland from the country between the Severn and the Marches of Wales, which, of course, is not far from the Cotswold hills. The long woolled sheep came from lower lands, the two best known long woolled breeds being the Cotswold and the Lincoln. By the fifteenth century these were, according to Eileen Power, the source of most of the fine wool. (See also the author's *The Wool Trade Past and Present* for a longer discussion of the subject.) Eileen Power's rather surprising statement suggests that Cotswold

wool was both fine and long, which appears contradictory, but we must be careful because we are not certain what medieval woolmen meant by fineness and length. The two have always been relative. Before attempting to decide about the medieval Cotswold it is best to consider what earlier writers about the sheep have said. Youatt in 1837 stated that the wool was long and coarse and he must be accepted for the period when he wrote; this was, however, after Bakewell's work had brought about the great change in British wool. Rather earlier, the accurate Gloucestershire historian Rudder, writing in 1779, stated that it was coarse. But earlier it is more difficult. Youatt attempted to argue that it was always coarse but I doubt if his attempt to prove this by statistics is sound. What else have we? Rather strangely only the poets, especially Michael Drayton, friend of Shakespeare and author of a few magnificent sonnets plus a long poem entitled *Poly-oblion* which lists all the many national attributes of England, among them the famous Cotswold wool:

T'whom Sarum plaine gives place,
    though famed for flocks
Yet hardly doth she tythe on Cotswold's
    wealthy locks
Though Lemster him exceed in fineness
    of her web
Yet quite he puts her down for his
    abundant store.

Although it is impossible to be certain, I think that medieval Cotswold sheep, which provided the basis for the fame of the Northleach and other British markets, produced short fine wool. Probably the coarsening had already begun with the Tudor enclosure movement. It was certainly completed by the coming of the New Leicester. As Luccock, the distinguished writer on sheep, wisely says, it is not likely that these lofty hills would have supported a heavy sheep at the time when agriculture was but little understood.

Devon Closewool

The Devon Closewool looks like a cross of the Devon Long Wool sheep with the Exmoor Horn, and this is the usual opinion. Several generations of careful selection fixed the breed which thrives well on the high lands or in the valleys. It is certainly a relative newcomer and is not mentioned in the older books on sheep.

Devon Longwoolled

The sheep of Devon and the problem raised as to their derivation is a microcosm of the national position. Today there are at least five clearly distinctive breeds. Two long woolled types, the Devon Longwool and the Devon Close, two hill types, the Dartmoor and the White Faced Dartmoor, and finally the very distinctive Exmoor Horn. The Devon Longwool is said to be a sheep of great antiquity but I have been unable to find much evidence for this statement. There is a good account of the sheep of Devon and Cornwall in Marshall's *Rural Economy of the West of England* but it is difficult to link the position then with the breeds we find today, except that then, as now, the wools were usually shorn without washing. The wool of the Devon Long is among the coarsest produced—certainly the coarsest in England, and was traditionally used for Long Ells, a major product of the Devon cloth industry and sold extensively to the East India Company. There has clearly been

a considerable addition of Leicester and possibly Lincoln, but it is difficult to account for the distinctive fleece of this sheep.

English Longwool (type)

The typical white-faced English sheep which in its modern form owes so much to Bakewell's Leicester and whose blood will be found in almost all the breeds detailed in this section.

Hexham Leicester See Bluefaced Leicester.

Kent

Alternative name for Romney Marsh, at least in England. When Romney Marsh is crossed, the resulting animal is usually called the Kent Half-bred.

Leicester

The English Leicester is now a comparatively rare animal but fifty years ago it was quite common in England. Its main interest is, however, historical, and in the story of sheep breeding there are few, if any, events as important as what Bakewell did to the Old Leicester sheep. It was a heavy, flat-sided, big boned sheep which during the years from 1760 onwards at Dishley, Bakewell changed into a new breed, establishing thereby an epoch in the history of the meat producing sheep, indeed of all sheep breeding. Bakewell adopted as his guiding maxim that a big-boned sheep was a poor sheep and he evolved a new type of animal which became known as the New Leicester or Leicester Shorthorn and, by inbreeding, established the light-boned, round animal, about as wide as it was long. Before he started his work the average full grown sheep produced 13 kg (28 lb) of mutton and a lamb of 8 kg (18 lb). The improved breed gave 36 kg

(80 lb) and 23 kg (50 lb) respectively. Its fame spread rapidly and Bakewell was soon letting out his rams for 50 guineas instead of a few shillings. His most famous ram was 'Two Pounder' which was sold in 1789 for 1,200 guineas. Although Bakewell's followers believed that the New Leicester was destined to replace all other breeds, this has not taken place. After having the great effect on many other British breeds already indicated, it has now declined in number as a pure breed and is comparatively rare, but this does not lessen the influence of its work. In practice today its great part in sheep breeding is more clearly shown in the Border and Bluefaced Leicester.

Lincoln
The Lincoln sheep today is a development of the old Lincoln which was changed and improved as a meat producing animal but not as a wool producer, by the New Leicester sheep. Arthur Young, in his *Agricultural Survey of Lincoln*, has a great deal of information on the struggle between the breeders who wanted to maintain the old Lincoln and those who had been captivated by what Bakewell was achieving. There is some difference of opinion about how the old Lincoln sheep developed. Gervais Markham in the sixteenth century wrote, 'Lincolns, especially on the salt marshes, have the largest sheep but not the best wool, for their legs and bellies are long and naked and their staple is coarser than any others'. But other writers have praised the wool and certainly it was, by 1800, the best long wool in the country. The improved Lincoln was more widely used for crossing abroad with the Merino than the

Leicester. It helped build the New Zealand crossbred trade and in Australia it played a part in making the Corriedale. Because of its continuing use in this way, there are now more Lincolns than Leicesters in Britain. Excellent, purebred Lincoln flocks are found widely spread through the main sheep producing countries and the best Lincoln wool I have ever seen was in a wool sale at Melbourne.

Romney
The Romney Marsh sheep has always been a native of the area that gives it the name, a low lying, bleak, exposed section of Kent interspersed with inlets from the sea—a countryside that will stay long in the memory of those who know it. The marsh was drained many years ago and now produces rich pasture, but the sheep remain and are raised on grass only. They mature early in spite of adverse weather conditions and they stand heavy stocking. The original sheep is said to have been improved by Leicester blood but it seems to have had little effect; few sheep, except perhaps the Soay, have been so long unchanged.

Reading the accounts of earlier writers it is clearly the same sheep we see today. For example, take an entry in the *Agricultural Society* of 1796: 'These sheep, called in the country Romney Marsh, but at Smithfield where great numbers are sold every week, Kent sheep, are remarkable for arriving at an extraordinary degree of maturity at an early age, and for producing a large fleece of long stapled wool of medium quality. These combined render it perhaps the most valuable sheep in the world'. Or take a quotation from that staunch Englishman, William Cobbett: 'The sheep are a breed that takes its name from the Marsh ... Very pretty and large ... with sheep such as I have spoken of before the Marsh abounds in every part of it, and the sight is beautiful'. E. Brindritt in an article on the Romney Sheep contributed to *Wool Husbandry* and reprinted in *Sheep Husbandry at Home and Abroad*, noted: 'Whenever one sees a flock of Romneys grazing, no matter how large or small the pasture or how few the sheep in number, they will be spread over the whole area evenly and on flat ground will remain so at night. This habit of spreading is what makes them so outstanding as grazers.'

The farmer always expects them to fend for themselves, providing no shelter even at lambing time. However, despite the very considerable interest shown in the Romney sheep in its native habitat, it is what happened in New Zealand that gives the breed its great place in sheep history, and that is described in a separate entry.

South Devon

A very similar sheep to the Devon Longwool but larger.

Wensleydale

The Wensleydale sheep takes its name from the beautiful Yorkshire dales where, in 1838, Mr Outhwaite of Appleton hired a large Leicester ram from a Mr Souly, a well known Leicester breeder, for the season. One of the progeny of this crossing with the older, local Teeswater breed was the famous 'Blue Cap' and this sheep was the foundation of the breed. According to the Society booklet, he possessed a wonderfully broad masculine head of a deep blue colour and his wool was fine and lustrous. Not only did Blue Cap and his

descendants stamp the Wensleydale with the wonderful lustre wool and bold carriage, but it strengthened the shoulders and neck and gave the breed the strong deep fore-ribs for which they are so well known. This appears to have been the only Leicester blood that was introduced.

As regards wool the Wensleydale has the most lustrous and curly wool ever grown, the fleece hanging in spiral ringlets. No woolman who has ever seen the sheep or wool will forget it. The heyday of the breed's popularity came around the turn of the century when it was found all over the lower lands of North Yorkshire and Durham. At that time the wool was sold in Bradford as Ripon hog, but despite the distinctive wool the sheep has always been of much more importance as a crossing sheep. The Greyface cross between the Wensleydale and the Blackface was widespread over the lower Pennines. Even more important, a Half-bred Wensleydale/Scottish Half-bred was known as the Masham (see separate entry). Recently, however, the Wensleydale has lost some of its popularity and has become comparatively rare, but it has played an important part in the founding of the Bluefaced Leicester.

Shortwool Breeds

These are mainly Down sheep, but include the Dorset Horn and Cheviot breeds.

Cheviot

The Cheviot sheep is one of the few real dual-purpose sheep, excellent for both meat and wool. In the eighteenth century it was confined to the Borders and was referred to as the 'long' sheep to distinguish it from the Blackface or Forest

Cheviot.

'short' sheep. Sir John Sinclair played an important part in the development of the breed, perhaps using some Merino and Southdown blood, and as a result the Cheviot became much more widespread, displacing the Blackface for a time, but during the early part of the twentieth century there was a reversal of this trend.

Another pioneer in the improvement of the breed was Mr James Robson of Belford, Roxburghshire a farmer on Bowment Water who, like Bakewell, travelled through England and Wales to find the sheep he wanted.

The wool has traditionally been the basis of the famous Scottish Tweed largely made in the same part of the south of Scotland, and although other wools are blended—traditionally Southdown and New Zealand Half-bred—the connection between wool and cloth is so close that the word Cheviot has become generally used for all wool and cloth other than that of Merino quality. The Cheviot sheep with its distinctive white face, produces a wool remarkably free of coloured (i.e. usually black) hairs and when light shades are required this is important, and adds to the other value of crispness which is so distinctive.

From a breeding point of view the Cheviot has yet another claim to importance, the ewe often, after having served a period on the higher ground, is crossed with the Border Leicester to give the so-called Half-bred which in itself is the basis for many English Down Cross lambs. This cross—and it should always be remembered that the majority of sheep seen in Britain today are crosses—has in the past been called 'the rent payer of the Borders'. The term might today be given a much wider designation, indeed in the writer's own lifetime in Wessex he has seen the old traditional Down breeds to a considerable extent replaced by white faced sheep, that is the Cheviot-Border Leicester cross or, if the Down sheep have retained a footing, it is by the Down/Half-bred cross.

Dorset Down
The origin of the Dorset Down sheep is not quite clear but it probably derived from the Hampshire Down improved with the Southdown. The difference between the Hampshire and Dorset Down sheep is minimal. Today its popularity has declined and it is rarely found outside the county that gave it its name. I knew this sheep or, rather, its wool better than any other British breed and have frequently bought it. It is very attractive and good handling and could be mixed with finer wool, even Merino, without risk of detection. It also made a very distinctive cloth when used alone, its only real fault lies in the presence of black hairs.

Dorset Horn
The Dorset Horn sheep is more important than its fellow countryman the Dorset Down, and is indeed a very distinctive breed both in its present state

Polled Dorset, USA.

Hampshire Down, USA

and historically. Probably its main importance comes from its capacity to produce lambs at any season, and for this reason it has been widely used to produce Christmas lambs. An early reference to this practice in 1757 is in Lisle, *Observations on Husbandry*. Rather like the Ryelands and unlike most other British breeds, it was not much affected by either the New Leicester or the Southdown. Perhaps with the Cheviot, which it a little resembles, the Dorset Horn is the best dual purpose (meat and wool) sheep that we have today. During the twentieth century a Polled Dorset has been developed which has proved particularly popular abroad.

Down (type)

The way in which the Down sheep of southern England was evolved by Mr Ellman has already been described. The main strains are the Southdown, Hampshire, Suffolk, Dorset, Oxford and Shropshire, all of importance in sheep breeding today.

Exmoor Horn

The Exmoor Horn is another interesting development of the old Western or Wiltshire Horn but, unlike the Dorset Horn, it has not been widely used outside the area that gave it its name.

Hampshire Down

The Hampshire Down sheep was established by 1815 in the counties of Hampshire, Wiltshire, Dorset and Berkshire by improving the local native sheep with some Southdown and a touch of the Horn sheep. One likely method was by crossing the old Wiltshire Horn sheep and the Berkshire Knot with the Southdown. The breed spread rapidly and in Wiltshire it replaced both the native Wiltshire Horn and the Wiltshire Down. Forty years ago there were many more Hampshire Down sheep in Wiltshire than in Hampshire but recently it has lost ground, both to the Half-bred (i.e. Cheviot/Border Leicester) crossed with Down (usually Suffolk) mainly because of its comparatively low lambing rate. It has, incidentally, a great deal of wool on its head which some have thought indicative of the bad lambing rate. It is a massive sheep and now more popular abroad than in Britain.

Hereford *See* Ryeland.

Leominster *See* Ryeland.

14   Devon Longwool.

15   Devon Closewool.

16   South Devon.

17   Dorset Down.

North Country Cheviot

Travelling through Scotland one comes first to the Cheviots on the hills of that name and then, having passed through the central industrial belt and reached the Highlands, it is the Blackface that takes over. Finally, having reached the delectable north west—Wester Ross and Sutherland—it is again the Cheviot. So much so is this the case that the Cheviots native to that area have become recognised as a breed in their own right—the North Country Cheviot—with a number of differences to the other, older Cheviot breed. Its arrival in north Scotland was mainly due to the noted Sir John Sinclair ('Agricultural Sir John' as he is often called) who did so much for animal breeding and was the first president of the Board of Agriculture. He is said to have given the sheep its name and he was one of several breeders who developed the Cheviot we know today from the old, so-called Long Hill sheep. He also took 500 ewes to his Caithness farm and ever since the Cheviot sheep has been the favourite breed in that part of the world, indeed, breeders there claim that they have stayed nearer to the original Cheviot than their southern rivals, but such a point is difficult to prove. It is enough to say that the North Country Cheviot produces excellent meat and a wool with all the qualities sought for when manufacturing Scottish Tweed. Today it is the more numerous of the two Cheviot breeds.

Oxford Down

The Oxford Down was originally a cross between the Hampshire Down and the Cotswold, as improved by the Leicester, showing clear signs of the Cotswold ancestry. Its success was a major reason for the decline of the pure Cotswold. It has been widely used for crossing, notably in that great sheep country, the Scottish Borders, where it was taken around 1867. Like the Hampshire Down it has a distinctive woolly face or rather, woolly brow, but it can be distinguished from the Hampshire where the wool comes well down the face, and from the Shropshire where it almost covers it.

Poll Dorset

This is a strain of the Dorset Horn with, as the name implies, no horns. It has been considerably used abroad.

Ryeland

The Ryeland sheep comes from Hereford and the breed, as indicated in the historical section, is important. In Tudor and Stuart times it was often called the Leominster after the main town in the area. Camden writes of the town: 'The greatest name and fame that it hath these days is of the wool in the territory round about it'. At that time the wool was described as equal to the web of the silkworm in fineness and for softness, to the maiden's cheek. Camden himself put it more prosaically: 'Lemster ore they call it, which, setting aside that of Apulia and Tarentium, all Europe counteth to be the very best'. And it was certainly changed less than most, by the Bakewell revolution.

Shropshire

The Shropshire sheep is now much less popular than it was; perhaps with the decline of the Down breeds generally it was to be expected that this breed would lose out as it came from a part of the country which is not typical for Down breeds. It was evolved from the old

Shropshire.

Morfe Common breed crossed with Southdown, and, as with other Down breeds, it has retained its position better abroad. Distinction is afforded to it as being the first to have published a Flock Book and it is to the credit of the breeders that a society was formed as early as 1882.

Southdown

The Southdown is one of the great sheep of history. Its early story is more or less wrapt in mystery but it is clear from early times, probably the pre-Roman, that upon the chalk downs that gave the sheep its name there was a native short woolled breed. This was the starting point for the work of John Ellman of Glynde and his own life is practically a history of the Southdown sheep. Arthur Young in his *Annals of Agriculture* has much interesting information regarding John Ellman and this sheep. As the account of the down sheep of England has shown, this sheep changed the story of farming on the chalk hills of southern England and, during the late nineteenth century, sheep based on the local breeds

crossed with the Southdown dominated the scene. Equally important, they were widely employed abroad, perhaps most notably in New Zealand. It is not too much to claim for this sheep that the Canterbury frozen lamb trade was built upon it and established by the use of the Southdown ram on the Romney Marsh and Leicester ewes. The lamb cross matures well, showing great quality.

Suffolk Down

Among the Down breeds today in the British Isles, the Suffolk is the only type to have improved its position. Originally from the eastern counties that gave it its name, it is now found on all the southern chalklands. Its ancestry is well known, being a cross made around 1800 between the Southdown as improved by Ellman and the Old Norfolk Horn, which had long been established there. The black faced, horned, or heath sheep has bequeathed much to the Suffolk Down, including its black face, and from the wool point of view a superfluity of black hairs which, again from the wool point of view, makes it probably the least desirable Down wool. But its present high position in sheep breeding depends on the use of the ram on the Half-bred to give both good lambing rates and good meat—the two main requisites of commercial sheep farming in Britain today.

Wiltshire Horn

This is a very different animal to the Dorset and Exmoor Horn breeds. It would seem likely that it was derived from the old Wiltshire Horn which was the distinctive sheep for folding in Wiltshire. (See historical section.) If this is the case, it shows what selection will do, because the Wiltshire Horn today has no

53

18 Oxford Down.

19 Ryeland.

20 Southdown.

wool and indeed looks rather more like a goat. It is used as a ram for getting good lambing rates and meat but has not been as widely adopted as the Border Leicester, Bluefaced Leicester and Suffolk Down for this purpose.

Hill and Mountain Breeds

Blackface Mountain (Type)

The Blackface sheep today is the most numerous breed in Britain. It is sometimes regarded mainly as a Scottish sheep but it originated in England and moved north to replace the smaller, more native Scottish sheep of which existing versions still remain in the Shetland and the Soay. (All, incidentally, smaller sheep with finer wool.) The Blackface went north and came, with the Cheviot, to dominate sheep farming in Scotland. Now it is coming south again and in the western parts of Britain (which have for many years, even centuries, had their own typical mountain breeds such as the Herdwick of the Lake District, the various breeds of the Welsh Mountains and those of Devon) the Blackface is everywhere rather taking over. Elsewhere in the world where conditions are hard it has been found an outstanding breed. This sheep has always maintained a certain wildness in its nature, which above all has made it the excellent hill sheep that it is. The meat is excellent, reminding one of Peacock's classic lines:

The valley sheep are fatter
But the mountain sheep are sweeter!
We therefore deemed it meeter
To carry off the latter!

Black Welsh Mountain

During the past eighty or a hundred years, this sheep has been developed by selecting black lambs from the Welsh Mountain breed and in this way a distinctive breed with its own Breed Society and Flock Book has been established.

Clun Forest

The Clun Forest today is the most numerous of the sheep of the Marches of Wales. Its origin is obscure and the wool, which is considerably better than that of the other sheep of the area, suggests a certain amount of Ryeland blood. Twenty years ago this sheep spread to other areas, notably the chalk downs of Southern England but it has hardly maintained its position there. Like the other sheep of the Welsh Marches, the Clun has made comparatively little mark on the world scene.

Dales-Bred

The Dales-Bred comes from Upper Wharfedale. The Breed Society was established in 1930 by a group of breeders there. The sheep was doubtlessly established by selective breeding from the local forest or heath sheep, with possibly some outside blood.

Dalesbred.

Dartmoor

The Dartmoor is one of the largest

moorland or hill sheep in Britain and in appearance rather resembles the Devon Longwool breed. The two have probably been interbred. The Dartmoor clearly derives from the native heath sheep improved by the Leicester, perhaps by way of the Devon Longwool. Vancouver, in his *General View of the Agriculture of the County of Devon*, has a description and a drawing of a Dartmoor sheep, c. 1800, which shows a somewhat different animal, presumably before the 'improving'.

Galway

The Galway is now the predominant lowland breed in Ireland and is found in the county from which it takes its name and other districts around. Earlier it was allied to the older Roscommon breed but split from it around 1926 and in the end absorbed the older breed. Its white, hornless head clearly indicates the basic Border Leicester blood.

Gritstone (Derbyshire Gritstone)

This is another Pennine breed of the basic Blackface Mountain type of very similar qualities to the Lonk, with which it is sometimes crossed. It was doubtlessly derived in the same way.

Herdwick

The Blackface which is strongly extending its position as the most popular British hill sheep, has made little progress in the Lake District, where the Herdwick holds supreme and, in a way, this breed can be reckoned the most distinctive of all English hill sheep. It is also one of the hardiest sheep and there are many stories of its survival under harsh Lakeland winter conditions. Entire flocks have been buried under fifteen feet of snow and have survived. The breed is

certainly one of the oldest in Britain and its name probably derives from two Saxon words, 'herd' meaning enclosure and 'wick' meaning village. But this of course does not mean that the sheep has not changed since Saxon times. The often quoted story of a Spanish origin can be dismissed for, as Dr Oliver said years ago, it probably displaced—or impoved by cross-breeding—an earlier forest breed.

Among the many individual habits of the sheep the most remarkable is its homing instinct. A Windermere farmer once sold part of his flock to another farmer in the Workington area of the Cumberland coast. Some days after his purchase, not a single sheep was to be found in their new abode. Another ewe was taken from the high hill east of Ullswater to the Skiddaw country beyond Keswick and swam back across Ullswater to her native home. As a result, Herdwick sheep are almost always sold with the farm and, if the farm is let, then they go with the holding, the tenant finding security for re-delivery at the end of the lease. Cloth made from the wool of the Herdwick sheep used to be called Hodden Grey and, as Burns wrote:

What tho' on hamely fare we dine,
Wear Hodden-grey and a' that,
Gie fools their silks and knaves their
    wine—
A man's a man, for a' that.

There is unlikely to be any great expansion of the Herdwick sheep; outside their own lakeland they are unlikely to replace the Blackface, but there—both because of their own virtues and because the National Trust of Britain is now one of the most extensive stockholders of the breed—they seem secure. It is good

21 Exmoor Horn.

22 Scottish Blackface.

23 Suffolk.

that this should be so for it would be wrong that such a distinctive breed should disappear. The Herdwick is remarkable for its long life and although lambing rates are not high, the great age which many ewes reach does enable them to produce a remarkable number of lambs. One Herdwick ewe at Matterdale, near Penrith, lived to over eighteen years and had thirty-three lambs, all of which lived.

Kerry Hill

The Kerry Hill is one of several hill sheep found in the Marches of Wales, and has been long established.

Lonk

Moving south from the country of the Herdwick, Swaledale, and its kindred sheep, into south Yorkshire and Derbyshire, there is the old established blackfaced breed—the Lonk. It is not mentioned as such in Youatt but the breed has clearly roamed the Pennines for many years and although there has doubtlessly been some interbreeding with the Blackface and the Swaledale, the breed as we see it now is mainly the result of adaptation to environment over many years.

Radnor

The Radnor is another breed of the Welsh Marches closely linked to the Kerry and perhaps deriving from a cross of that sheep with the Welsh Mountain.

Rough Fell

The Rough Fell is one of the lesser known hill sheep found almost exclusively on the High Shap Fell area. The Breed Society dates from 1926 but an earlier reference to this sheep will be found in Garnett's *Westmorland Agriculture 1800–1900* which stated there were then in the county three breeds—the Herdwick, the Swaledale and the Rough Fell.

Scottish Blackface

By far the most important blackface sheep in Britain today, it is much used for crossing, usually with the Border Leicester, and has two quite distinctive roles in meat production, firstly in the lambs it produces on the hills and secondly, the lambs it produces when brought down later to the lower country and crossed with the Border Leicester.

As far as the wool is concerned it has many uses, indeed a surprising variety of them. For example, it is excellent for stuffing mattresses and upholstery. There has always been a keen Italian demand for this purpose. Italian girls buy the fleeces for their bottom drawer and avoid any with black fibres, since these are thought to make their marriage infertile. (Ryder: *Scottish Sheep and Wool*). It must be quite a job! The wool is also good for carpets, ranking second only to East Indian, which is best of all for this purpose. Finally there is cloth, the best known being the home-woven Harris Tweed of the Scottish Isles, both the best and the finest as well as the coarse being used, depending for their relative popularity on fashion demand.

The origin of the breed is not known for certain; there is a tradition that the sheep came from a Spanish ship wrecked during the northward flight of the Armada in 1588. The same story has also been used to explain the presence of the Herdwick sheep in the Lake District but in neither case need it be given any credence. No Spanish sheep has a particularly close resemblance to the Blackface or the Herdwick. The answer to the

Blackface sheep in typical habitat.

question of origin must be sought elsewhere and doubtlessly in England, perhaps on the Pennine Hills. As far as

Scotland is concerned, Hogg, the Ettrick shepherd and poet, stated that in 1503 James IV established a flock of 5,000 in Ettrick Forest. Dr Thomas Oliver, who knew the area well, wrote in his book *Wool* published fifty years ago, that 'of late severe winters have decided grazers to favour the hardier Blackface rather than the Cheviot. They fatten well on poor feed. The breed has spread into the Highlands, the Cheviot Hills, and South West Scotland, along the Pennines to Derbyshire and to Northern Ireland. So that now about half the Scottish sheep are Blackface and its Border Leicester Greyface cross'. This development has continued.

South Wales Mountain

A larger version of the Welsh Mountain sheep found in Glamorgan and Monmouth. It has its own Breed Society.

Swaledale

Unfortunately, comparatively little is known about the history of the Swaledale breed, which is becoming increasingly popular. The Swaledale Sheep Breeders Association was first registered in 1919 and hardly any of the history of this breed was recorded before then.

Obviously, however, the hill farmers of the border country of North Yorkshire have for many generations been breeding sheep suitable to the environment and based upon a careful selection of the progeny. As with so many British breeds it is this strong local acclimatisation to the area that is so interesting. Today the Swaledale is much used for crossing, giving, with the Leicester, Border Leicester or Blueface Leicester, the Greyface or Mule, and with the Teeswater or Wensleydale the well known Masham.

Teeswater

The Teeswater is indigenous, as its name implies, to Teesdale and for the last 150 years has been bred in that area to provide rams for crossing with hill ewes—either Swaledale, Dales-bred, Rough Fell, Blackface, Herdwick, Cheviot and North Country Cheviot—to produce the well known half bred, the Masham. Although it produces useful meat and wool, the sheep is almost entirely judged on its capacity as a sire.

Welsh Mountain

The Welsh Mountain is really a generic name covering several more or less distinctive breeds. There are, in fact, seven breed societies as follows:

Welsh Mountain—Hill Flock Section

24 Teeswater.

25 Swaledale.

26 North Country Cheviot.

Welsh Mountain—Pedigree Section
Welsh Mountain Rams
Black Welsh Mountain
South Wales Mountain
Beulah Speckled Face
Welsh Half Bred

Welsh wool has always been rather unkindly treated, thus Markham in *Cheape and Good Husbandry* (1614) wrote: 'The Welsh sheep are of all the worst, for they are both brittle and of the worst staple and indeed are praised only in the dish for they are the sweetest mutton'. An interesting statement, which indicates the importance of wool rather than meat when Markham wrote. This of course has now completely changed and it is the meat that matters, and is indeed very good. The wool, however, does show great variety.

White Face Dartmoor
The White Face Dartmoor was, according to the Society which was formed in 1950, at one time more widespread than it is now, but historical evidence for this is rather lacking.

Wicklow Cheviot
This is now usually described as a strain of the North Country Cheviot and the Breed Society only dates from 1943, but there was an older Wicklow sheep mentioned and illustrated in D. Low, *On the Domesticated Animals of the British Isles* 2 vols (1842).

Rare Breeds

Badger-faced
This is the least known of the Welsh Mountain sheep types and is now extremely rare.

Beulah Speckled Face
A speckled face hill sheep bred for over

a hundred years on the hills of Eppynt, Llanafan, Aberywesyn and Llanwrtyd Wells, now has its own breed society.

Cannock Chase
One of several breeds native of the Welsh Marches with the Longmynd and Morfe Down all of which played an important part in the formation of the Shropshire Down sheep.

Cardy
A specialised variation of the Welsh Mountain.

Eppynt *See* Beulah Speckled Face.

Hebridean
Alternative and probably better name for the St Kilda.

Jacob
The Jacob is today hardly a rare breed; it is very popular as an animal for parks, etc., and the wool is much liked by home spinners. It is a rather mysterious animal but there is probably nothing in the biblical origin story. The sheep appears to be of Scandinavian origin and linked to such breeds as the Manx Loghtan.

Lleyn
A local North Wales variety of the Welsh Mountain.

Llanwenog
Somewhat similar to the Clun, the basic sheep traditional to parts of West Wales, it has come into some prominence this century when it has been improved by crossing with the Shropshire and there is now a Llanwenog sheep society.

Longmynd
A blackfaced type sheep which, according to Trow Smith, became extinct when a flock died out at Longmead

Farm, Craven Arms in 1926. It probably played some part in the formation of the Shropshire Down.

Manx Loghtan

This is one of the largest and rarest of the distinctive Scandinavian-derived sheep. It is found on the Isle of Man and has a particularly fine set of multiple horns.

Morfe

This is another sheep coming from the same district as the Longmynd. It probably played as important a part in the foundation of the Shropshire as did the Longmynd. As far as the present writer is aware, the breed is now extinct.

Norfolk Horn

The Norfolk Horn sheep were once amongst the most numerous in England and their wool was highly regarded. Un-fortunately by 1969 there were only three rams and three ewes still remaining and the resultant progeny showed great weakness, but by crossing with the Suffolk the type and potential of the old breed is being retained in the New Norfolk Horn.

North Ronaldsay

Until recently the North Ronaldsay or Orkney breed was only on the island of North Ronaldsay where the sheep graze on the shore, feeding almost entirely on seaweed, and being prevented from getting to the richer, inner land by a wall that has been built around the island. The Rare Breeds Survival Trust, who are doing such excellent work with these rare breeds of sheep, have now taken a flock to a new home, the island of Linga Holm, so as to safeguard its survival.

Rhiw Flock (now very rare).

27 Welsh Mountain.

28 Herdwick.

29 Dorset Horn.

Orkney See North Ronaldsay.

Penistone
The Penistone is another Pennine breed, much mixed with the Lonk and Derbyshire Gritstone and derived in much the same way. It is now almost extinct and does not appear in *British Sheep* 1976 revised edition. It produced a better wool than most of the rare breeds mentioned here and was probably derived from a different background, not Scandinavian but the original Soay.

Portland
An old breed, native to that part of Dorset from which it gets its name. There are only a few of these sheep in existence today.

Rhiw Hill
A local variant of the Welsh Mountain breed, found in South Caernarvon.

St Kilda
The St Kilda is a small, multi-horned sheep which is found on the island of that name and must not be confused with the Soay. It has been suggested that it would be better to give this breed the name of Hebridean. It is clearly of Scandinavian origin.

Shetland
The Shetland sheep is very much a native of the islands and is, indeed, hardly found elsewhere. Almost certainly of Scandinavian origin it has probably not greatly changed during the past hundred years. Although the Cheviot and the Blackface have been taken to the islands and there are a number of flocks of the latter, crossing would seem to have made much less difference to the native Shetland sheep than one might have expected.

The fleece is still light in weight, often not more than 2 lb which was the average over the whole of Britain two or three hundred years ago. Until recently it was still plucked or rooed, but is now usually shorn, and traditionally goes to the making of knitwear for which the Islands have long been famous. The wool is of several colours and is used in its natural shade for the traditional knitting. Moorit (mid-brown) and Skiela (brown-grey) are particularly distinctive. The sheep were fed on the scathold or common land near the sea but now they seem to be spreading almost all over the islands, although of course no part of the Shetlands is far from the sea. The wool is particularly soft handling and when one considers the relative coarseness of the actual fibre, this is quite remarkable. The wool is also rather mixed, the Shetland being one of the few sheep in Britain today where the old traditional coarse outer and fine inner fibres can be distinguished. An attempt has been made recently to identify separate strains of the Shetland sheep but this is perhaps not necessary as my own knowledge of the sheep on the Islands suggests that it remains a remarkably homogenous breed.

Soay
As stated earlier in this chapter, the Soay sheep can be reckoned the only living representative of the small primitive sheep that was common in Britain before the Roman occupation. They are much older than, for example, the Shetland, which was probably introduced to the northern islands by the Scandinavian conquerors around A.D. 500. The Soay is a most distinctive animal, small, with good legs, carrying a short brown fleece

which it sheds naturally in the early summer.

Today a few flocks of Soay can be found in England in parks and rare animal museums but only on one of the St Kilda islands (Hirta) can they be found in their natural habitat. How they got there, and indeed when man first visited this group of distant islands is not known. It is in fact a little surprising that it should be the Soay sheep rather than the Shetland that is found there. Hirta, the main St Kilda island, was inhabited continuously for many years before being evacuated in 1930. In 1932, 107 sheep were brought from the neighbouring island of Soay and released on Hirta and by 1952, when the new count was taken, they had increased to 1,114. The evacuation of Hirta has given scientists a rare opportunity of studying in detail what would be impractical elsewhere, namely what happens when a group of sheep are left to run wild with no control whatsoever by man. Among the interesting information the scientists working on Hirta have discovered, is that there is on average one ram for every five ewes. As obviously there must be approximately the same number of ram and ewe lambs born, this is a rather surprising difference which must arise from completely different death rates, which it has been suggested are caused by the strain of the tupping period. The rams may be less well prepared for the bad winter conditions after their earlier hard work. The ewes, although they carry a lamb, or perhaps because of this, tend to survive better.

It is worth noting that the Soay wool is surprisingly fine and when compared with the real Mouflon, the shorter inner fleece has been clearly developed and bred so that the coarser outer one can only be found with some difficulty. The Soay therefore represents a very definite survival of a properly domesticated Neolithic breed. It is doubtful if there is any other animal that shows quite so clearly how much the Neolithic animal breeders had in fact achieved. This removal of the coarse outer hair should be stressed as it has been carried much farther than in many other sheep existing today; for example, in the Shetlands or even more noticeably, in two fleeced Asiatic breeds.

Whitefaced Woodland   *See* Penistone.

Recognised Crossbreds

British sheep today can be divided into two groups, the pure-breds already listed and the crosses detailed below. The latter are sometimes called 'commercial flocks', a good name in so far that it exactly describes their purpose. Two or more pure breds are combined to obtain the ideal sheep for producing lambs that fatten quickly for the meat trade. It is particularly interesting that the three groups of British sheep, the Longwool, the Down and the Hill and Mountain, are usually combined to achieve this result. The Longwool ram (more often either the Bluefaced or Border Leicester) is put to the Cheviot ewe (now more often the North Country Cheviot) or, alternatively, one of the Blackface sheep, to produce one of the standard crosses listed below. Then frequently, but not always, the resulting ewe cross is mated with a Down ram, usually the Suffolk. Breeds that do not fit into this system are tending to become ever more rare.

Masham
A very important cross produced by

30 Clun Forest.

31 Kerry Hill.

32 Hampshire Down.

Teeswater ram on Dales-bred, Swaledale or Rough Fell ewe. Recently two slightly specialised developments may be noted: Teeswater rams have been used on Blackface ewes in Scotland to produce a so-called 'Scottish' Masham, and on Welsh Mountain and Speckled Face ewes in Wales to produce the 'Welsh' Masham. The Masham is certainly one of the most important of the cross breeds at present so common in the British sheep industry.

Mule

The Mule is sometimes regarded as an alternative name for the Scottish Greyface. This is probably not quite true and it should be regarded as a North England variety of the Scottish Greyface, with the not unimportant difference that it is now usually obtained by using the Blueface Leicester, not the Border Leicester, ram on the Swaledale. Other breeds of hill sheep can be used but the cross indicated above is the most common. The Mule ewe can, as with other breeds of this type, itself be crossed with a Down ram to obtain excellent lamb carcasses.

Scottish Greyface

This is obtained by crossing the Border Leicester ram on the Blackface ewe. It is found in large numbers in Scotland where conditions are harsher and therefore make this cross more successful than the otherwise more common Scottish Half-bred.

Scottish Half-bred

Often called the 'rentpayer' of the Borders, now spreading widely into Southern England, the Half-bred has in the past been produced by crossing the Border Leicester ram on the Cheviot ewe. The Blueface Leicester is also used today. The resultant ewe can itself be crossed with a Suffolk ram to give a further cross that is very popular because of its early lamb production.

Welsh Half-bred

May be considered the Welsh equivalent of the Scottish Half-bred, here the Border Leicester or possibly the Blueface Leicester is used with the Welsh Mountain ewe, giving the most common of all Welsh sheep. There is a good lambing rate and excellent meat results. As with the other cases, further crossing can be done by using the Welsh Half-bred ewe with a Down Ram, producing equally good meat of perhaps rather more desirable formation.

Major Developments from British Breeds

American Leicester

The American Leicester is derived from the Leicester and the Border Leicester, probably more of the latter, and the Breed Society was established in 1888, but the sheep is not now as popular in America as it once was.

Drysdale

This is a coarse woolled sheep developed in New Zealand by selection of monogenetic variants of the New Zealand Romney between 1929 and 1967. Although a useful breed, neither the Drysdale nor any other similar development has really shaken the hold that the pure Romney sheep has on the New Zealand agricultural scene.

New Zealand Romney

It is impossible to overstress the importance of the Romney sheep in New Zealand. The breed dominates sheep farming there, as a result except in the Mackenzie country (i.e. the central part of the South Island) something like 90% of the sheep one sees are Romney or Romney crosses.

The New Zealand Romney sheep derives of course from the English Romney but there are many variations in the way it has been used in New Zealand. Although, as stated previously, the Southdown was originally crossed with the Merino, later and in other districts of the central part of the South Island it was also used to cross with the Romney and much of the so-called Canterbury lamb of today is a Romney/Southdown cross and comes now more from the North Island than the old traditional South Island. The New Zealand farmer uses the Southdown ram differently to the English farmer, who would not, during the present century, have used this ram in any combination of breed whatsoever to obtain his fat lambs. Nevertheless, the Southdown ram on the Romney ewe has been found to give a relatively small lamb of good quality, excellent 'package meat' as it has been well described. However, recent developments have suggested that the Southdown used in this sense will not remain as popular because of its comparatively low fertility. Where the Romney is crossed, it is more likely to be with the Border Leicester or the Cheviot, and both are being used in New Zealand. Even more important, however, and more likely to dominate the sheep scene, is the continuing use of the Romney as a pure bred sheep. The complicated crossing system that evolved in Britain does not appear to be going to come to New Zealand. As compared with the English Romney the New Zealand breed is rather smaller and the wool somewhat coarser, perhaps due to the amount of Lincoln blood that was bred into the New Zealand Romney during the early days of sheep farming there. The reason for the use of the Lincoln then was that the much straighter wool that the Lincoln produces meant that when it was a question of using the sheep on uncleared land, less trouble was met with the wool becoming entangled with the scrub.

In addition to the standard Romney, another type which perhaps should be considered as a separate breed, has been developed, called the N-Romney. This produces a coarser wool and at some periods this was found to be an advantage.

Some years ago in the 1960s, coarse carpet wools of 44s (38 microns) quality and lower were in great demand and were fetching better prices than the finer types. Scientists working at Massey College, one of the two main experimental sheep breeding stations in New Zealand found that the dominant gene—the so-called N-gene—gave the coarse wool. It also gives a mixture of relatively coarse and fine which, although unpopular with the makers of yarn for apparel purposes, is wanted for carpet yarns. This tendency to give the old familiar coarse hairs used to be a complaint with the Romney but for carpet purposes is so no longer.

In some cases in order to obtain full advantage of these special qualities, six month shearing is adopted. The Romney normally gives a 17 cm (7 in) wool, but the carpet trade only requires 8-9 cm (3-4 in) and this demand has led to double shearing, one a little before lambing and the other a little before the ram is put with the ewe.

33　Rough Fell.

34　Lonk.

35 Soay.

It is interesting to note that the work done on sheep fertility in New Zealand has suggested that it is the sheep with less wool on its face that produces the most lambs. One might have expected this from historical reasons in England, the Border Leicester being the classic example of a sheep with little wool on its face and yielding high fertility rates. Attempts have been made to link this with genetical factors but as a distinguished wool scientist once said to the present author, it may simply be that the ram can see better what it is doing when it is put to the ewe!

Despite the interest in carpet wool at certain periods it must always be remembered that it is the meat-producing side of New Zealand farming that is important, and for this the Romney has proved the outstanding sheep.

New Zealand Southdown
When, with the coming of meat refrigeration, New Zealand turned from being mainly a Merino wool producing country to a meat producer, the necessary sheep were imported from England. Three breeds played an important part, the Lincoln (which was the least important) the Southdown and the Romney. The Southdown was used mainly to cross with the Merino, or later the Romney, to produce the so-called Canterbury lamb. Elsewhere the Romney in a pure bred form, was found to be economically more successful.

Perendale
Another New Zealand development, in the years 1938-60 (but chiefly since 1947) is the crossing of the Cheviot with the Romney. There has been a Breed Society since 1960. The name derives from Professor G. Peren of Massey Agricultural College. Although the breed does not seem to have had very great success, I have had the opportunity of examining the wool and it is one of the nicest Cheviot type wools. Obviously the Cheviot sheep has passed its typical qualities forward. But of course, it is as a meat producing sheep that the breed has mainly been developed.

Historical

Bampton Nott
Appears frequently in early agricultural books, but today all we really know about it is that it was apparently the early ancestor of the Devon Long, and through the Devon Long, of the Devon Close, sheep.

Berkshire Nott
Generally speaking, the sheep, that have become extinct, were mainly small and fine woolled—proof if any was wanted that it was not fine wool that the late eighteenth century improvers were seeking. Among this type of sheep the Berkshire Nott is one of the more mysterious, for comparatively little is known about its appearance. We know that Ellman is said to have introduced some of this breed when he was endeavouring to improve Southdown, although the purpose of this does not appear to be very clear. West Berkshire must have been ideal sheep country; motoring for example around Lambourn, or coming down to the valley at Wantage, the visitor will be continually impressed by this fact. Here surely must have been one of the breeding grounds of the fine woolled sheep that produced the raw material of the well known West Country broadcloth, in the history of which the clothiers of Newbury—Jack o'Newbury, for ex-

ample, and Dolman who built Shaw House—played a part.

Foulas

Probably an alternative name for the St. Kilda breed.

Kerry (Irish)

Low illustrated this breed in *On the Domestic Animals of the British Isles*. He describes it as being small, with crooked horns, that were often absent in the ewe. There was a tendency for black or brown animals and his description of the fleece as being coarse and hairy in the back, but short and fine in the side, suggests a mane like the Soay rams.

Mendip

This was an interesting sheep that is now extinct. It came, of course, from the distinctive hills of that name and would appear to have been a fairly direct descendant of the typical, primitive Soay type wool sheep common in England. There are some interesting records of attempts being made to cross this native breed with the Merino but, as with other similar cases, nothing came of it. I suspect that the wool from the Mendip sheep was always rather mixed. There may even have been a little Scandinavian blood. Viking influence spread down to this part of the world, as is clearly shown by the names of the two islands off the Somerset coast, Steep Holm and Flat Holm.

Nottingham

The Lincoln breed reared in Nottinghamshire—the wool would appear to be very little different from the long wool of the Lincoln.

Old Lincoln

From all the breeds that were destroyed by the Bakewell Leicester, it is perhaps the loss of the Old Lincoln that is most to be regretted. It produced a most distinctive long wool. J. A. Perkins in *Sheep Farming in Eighteenth and Nineteenth Century Lincolnshire* (No. 4 Occasional Papers in Lincolnshire History and Archaeology 1977) gives a fascinating account of this breed and of the struggle it made to defend itself from the invasion of Bakewell's New Leicester. Bakewell, incidentally, particularly hated the old original Lincoln and once, for the amusement of himself and his fellow practical farmers from Leicester, bought some particularly ugly specimens as a standard of comparison with his own New Leicester. Bakewell did not appreciate the distinctive quality of the long wool. Incidentally, the modern or improved Lincoln which, as stated earlier, is this Old Lincoln sheep improved by Bakewell, still produces the best English long wool and has, I think, always tended to revert a little to the Old Lincoln. I am reasonably certain that the superb long wool I saw some years ago in the wool sales at Melbourne represented a reversal back to this quality.

Roscommon

This at the turn of the century was the most typical breed of the Irish Republic. There is an interesting account of it by R. O. Pringle in F. Coleman, *Cattle, Sheep, etc. of Britain (1887)*. The wool was of 40/46s quality but the meat was more important. As already explained, this sheep lost its identity with the development of the modern Galway. As it existed in the early part of the nineteenth century it clearly owed a good deal to Border Leicester blood. Culley, the man responsible for developing the Border

36 Derbyshire Gritstone.

37 Dartmoor.

38 Wiltshire Horn.

39 Black Welsh Mountain.

Leicester, visited Ireland earlier and said that he had never seen such ugly, ill-formed sheep as found there.

Silverdale

This breed is now extinct. It was at one time a breed of the Northern Pennines.

West Country Down

Alternative name for the Dorset Down, and in some ways a better name as the Dorset Down sheep in its heyday spread well outside the county boundaries.

Western Horn See Wiltshire Horn (Old).

Wicklow

There exists today, and as already listed under Hill and Mountain Breeds, a Wicklow Cheviot or Wicklow Mountain breed but this seems to have been derived almost entirely from the North Country Cheviot and cannot be identical with the Wicklow sheep which Low describes.

Wiltshire Down

This sheep probably followed the Old Wiltshire Horn on the chalk hills of Wessex but it was so close and so inter-bred with the Southdown that even those who have studied the development of sheep on the chalk hills find identification difficult. It was replaced by the Hampshire Down, which in the latter years of the nineteenth and early years of the twentieth century was the most typical (with the Dorset Down) sheep of the area.

Wiltshire Horn (Old)

This sheep although now gone, deserves special mention because for many years it gave the finest wool of all British sheep and was also vitally important for another reason. It stood out above all else for its capacity to walk long distances over the Downs during the day and then to be hurdled or folded at night so that the ground was well manured and made fit for corn. Defoe has a good account of this type of farming. The standard account of this sheep comes in T. Davis' excellent book *A General View of the Agriculture of Wiltshire* (1811). He writes: 'The sheep stock of this district is an object of the greatest importance. It may indeed be called the basis of Wiltshire Down husbandry.' He continues: 'The peculiar aptitude of the soil and climate to sheep, the singular use of sheep folding on arable land naturally light and loose, the necessity of making sheep the carrier of dung in situations where the distance from home and the steepness of the hills almost preclude the possibility of carrying it by any other mode and particularly the disadvantages that art has given the farmers of this district of getting early grass by means of their numerous water meadows whereby they are able to breed lambs both for the supply of their stock and for the market, are the principal reasons which have contributed to give to Wiltshire the high rank it bears among sheep breeding countries.' He stresses again that the sheep is kept for two reasons and concludes: 'The first and principally, is undoubtedly the dung of the sheep fold and the second is the wool. The improvement of the carcase was not hitherto thought a primary object.'

The present-day Wiltshire Horn which has no wool on it at all must have been bred from this sheep because in formation they are much alike. Unfortunately, it has not been possible to trace who was responsible for this.

4 OTHER BREEDS

The Merino and British breeds are reasonably homogenous types and recognisable as such, but with others this is less the case and as indicated earlier, a geographical basis for listing has been chosen.

In the broadest possible terms the position today can be best understood by remembering the varied, primitive types of sheep and considering how they were developed under differing environmental conditions. And next, seeing how these types were themselves further developed by the inter-breeding of the Merino and British breeds already described. Specialisation along these lines has gone further in some of the geographical areas than others, most notably in the USSR. Elsewhere in Asia the basic sheep type is the indigenous fat-tailed which produces a wool of mixed fineness best suited for carpets. Many attempts are being made to improve these sheep by cross breeding with Merino or British sheep with varying success. Commercially, however, Asia remains the main source for carpet wool. These sheep have other uses—meat, dairy and skin—but almost entirely for local consumption. The same overall picture is true of India and Africa, except for South Africa which is a major Merino-producing country.

Europe and America are quite different. In the industrial parts of Europe the sheep, as always happens, have decreased; elsewhere, notably in the less accessible parts of the Mediterranean chiefly upon the islands of that sea, sheep of many varied and fascinating individuality can be found. They will de-light the historian of sheep breeds as well as the naturalist but their importance today is not very great. In the USA the sheep population has been falling fast, from around 56 million in 1942 to around 14 million now and non-Merino and non-British breeds are few. In South America the decrease has been much less but the breeds are even more closely based on the Merino and the British than anywhere else. Something more about developments there will be given in the introduction to that part of this section.

USSR

The USSR stands out as the most important area in the world today both for new sheep breeding developments and increased sheep numbers. Unfortunately, information is somewhat lacking but I have been given access to a valuable report *US Sheep Breeding Team Visit to USSR* (September 12 to October 6, 1975) and this, combined with sources indicated in the bibliography, especially Mason's *A Dictionary of Livestock Breeds*, has I hope enabled me to give a summary of what has been happening. As regards numbers, sheep in the USSR have increased from 134 million in 1961-5 to 143 million in 1974. During this period world figures rose from 1,006 million to 1,032 million. In terms of wool the increase is more noticeable, USSR production rose from 361,520 thousand pounds to 461,000 whereas the world figure stayed almost unchanged, declining from 2,597,378 thousand pounds to 2,548,161.

The US report divides the main breeds into three groups, meat-wool, wool-meat, and fur (skin) according to which raw material is the main priority.

It is particularly interesting to note that a good proportion of the wool-meat breeds are grown to produce coarse wool, although it is not quite clear whether this coarse wool is the recognised Asian or East Indian carpet type with a variety of fibre qualities, or the specialised longwool of which the English Lincoln was the prototype. I have made an attempt, based mainly on the derivation of the breeds, to indicate which appears likely in each case. It is probable that the main emphasis has been on the former, and that many of these are, in fact, the unimproved sheep of the fat-tailed type mentioned in the first paragraphs of this section. There is a third specialised sheep produced in the USSR mainly for the fur or skin of the young lamb. This sheep, usually known in the world market as the Karakul and producing the Black Persian Lamb coat, seems to have been more thoroughly developed in Russia than elsewhere and there are several important varieties.

USSR Meat-Wool Breeds

Akhangaran
Area: Uzbekistan. Products: Meat and semi-coarse wool.

Aktyubin
Area: Kazakhstan. Products: Meat and semi-coarse wool.

Alai Fat-Rumped
Area: South Kirgizia. Products: Meat and wool, also pelts. Kirghiz Fat-rumped × with Precoce rams, (1934-53) later Saraja.

Altai Mountain
Area: Siberia. Products: Meat and medium wool. Tsigai × (Merino × [Merino × local coarse woolled]) 1940-5.

Azov Tsigai
Area: Ukraine. Products: Meat and medium wool. British Romney × Tsigai

Balbas
Area: Armenia and Nakhichevan. Products: Meat, coarse wool and dairy produce. Caucasian fat-tailed type.

Bozakh
Area: Armenia and Azerbaijan. Products: Meat, coarse wool and dairy produce.

Buryat
Area: Buryat Republic and Chita Province, Siberia. Products: Meat and coarse wool.

Carpathian Mountain
Area: South West Ukraine. Products: Meat and coarse wool. Derived from Voloshian and of the Zackel type.

Caucasian Fat-tailed (Type)
Products: Meat, coarse wool and dairy products. An original source of many Russian breeds.

Cherkasy
Area: North West Kuibyshev. Products: Meat, semi and fine wool. Variety of Russian long-tailed with possibly some English and/or Merino blood.

Chita
Area: Buryat-Mongolia. Products: Meat and wool. Developed at Voroshilov State Farm from Buryat × Finewool.

Chuisk
Area: Dzhambul. Products: Meat, semi-fine wool.

Chuntuk
Area: Ukraine and Crimea. Products:

Meat and coarse wool. Similar to Kalmyk.

Dagestan Mountain
Area: Dagestan. Products: Meat, semi-fine wool. Caucasian fat-tailed with Württemberg Merino × Gunib 1926-50.

Darvaz Mountain
Area: East Tajikistan. Products: Meat, semi-fine wool. Darvas improved by Württemberg Merino 1941-8 and since by the Caucasian.

Degeres
Area: North Caucasus and Kazakhstan. Products: Meat, semi-fine wool. Shropshire × Kazakh fat-rumped.

Estonian Darkheaded
Area: Central and North Estonia. Products: Meat, semifine wool. Shropshire × local short woolled type 1940.

Estonian Whiteheaded
Area: South Estonia. Products: Meat, semi-coarse wool. British Cheviot × local coarse woolled sheep.

Fat-tailed (Type)
Widespread. Products: Meat, coarse wool. Of great importance, there are many varieties.

Gala
Area: Aksheron. Products: Meat and coarse wool. Parent breed the Shirvan and the Balbas.

Georgian Finewool Fat-tailed
Area: Georgia. Products: Meat and fine wool. Soviet Merino × Caucasian × Tushin (local coarse woolled type) 1932-49.

Georgian Semi-finewool
Area: East Georgia. Products: Meat and wool. Rambouillet or Précoce × Tushin 1931-49.

Gorki
Area: Gorki, Volga River and West Russia. Products: Meat, semi-fine wool. Hampshire Down × Northern short-tailed, coarse woolled local sheep 1936-49.

Hissar
Area: Tajikistan. Products: Meat and coarse coloured wool. Direct descendant of local fat-tailed type. 490,000 in 1969.

Kalinin
Area: Moscow. Products: Meat and long wool. British Lincoln × Northern short-tailed c.1935.

Karabakh
Area: Azerbaijan. Products: Meat and coarse wool.

Kara-Kalpak
Area: Uzbek S.S.R. Products: Meat and coarse wool.

Kargalin
Area: Aktynbirsk, Kazakhstan. Products: Meat and coarse wool. Degeres or Saraja × Kazakh fat-rumped.

Kazakh Arkhar-Merino
Area: Southern mountain area of Kazakhstan. Products: Meat and fine wool. . Akhar Mountain rams with Novocaucasus ewes × Rambouillet and Précoce Merinos 1935-50. (686,000 in 1969.)

Kazakh
Area: Kazakhstan. Products: Meat and medium wool. Précoce × Tsigai, the Lincoln, Stavropol and North Caucasus mutton-wool.

Kuchugury

Area: Nizhnedevitsk, Voronezh. Products: Meat and coarse wool. Voloshian × Russian long-tailed.

Kuibyshev

Area: Kuibyshev and Ulyarov, Volga. Products: Meat and medium white wool. Romney Marsh × Cherkassy with back crosses to Romney Marsh and local coarse wool sheep or the Volshian. 1936–48. (364,000 in 1969.)

Latvian Darkheaded

Area: Latvia. Products: Meat and medium wool. From Shropshire, Oxford and German black-headed mutton × local Northern short-tail 1937.

Lezgian

Area: South west Dagestan and North West Azerbaijan. Products: Meat and coarse wool. Caucasian fat-tailed type.

Lincoln (Russian)

Products: Meat and long, white wool. Famous English breed important in Russia. (299,000 in 1969.)

Liski

Area: Voronezh. Products: Meat and long wool. Lincoln × Mikhnov back crossed to a Lincoln ram 1962.

Lithuanian Blackheaded

Area: Lithuania. Products: Meat and medium wool. German blackheaded mutton and Shropshire × local Northern short-tailed 1923–34.

Mazekh

Area: Armenia. Products: Meat and coarse wool, also dairy. Caucasian fat-tailed and similar to Balbas.

Mikhnov

Area: Voronezh. Products: Meat and coarse carpet wool. Long established.

Russian long-tailed native type.

Niznedevitsk

Products: Meat and coarse wool. Lincoln × Kuchugury. Similar to Liski.

North Caucasus

Area: Stavropol region. Products: Meat and medium white wool. Fine-woolled Stavropol, Lincoln and Romney Marsh 1940–60. (1,142,000 in 1969.)

Omsk

Area: Siberia. Products: Meat and medium wool.

Oparino

Area: North West Kirov. Products: Meat and medium wool. English mutton breed × local Russian long-tailed.

Précoce (Russia)

Area: Bellyo-Russia and Siberia. Products: Meat and fine white wool. Originally from France, developed in nineteenth century Russia into a distinctive breed with Spanish Merino, Rambouillet and possibly Leicester blood. Now an important sheep. (5,738,500 in 1969.)

Priazov-Tsigai

Area: Ukraine. Products: Meat and medium wool. Developed at Stud Farm Rosa Luxemburg by crossing Romney Marsh with Tsigai. Probably same as Azov Tsigai.

Russian Long-tailed

Area: South European Russia. Products: Meat and coarse wool. From old Russian thin-tailed. Probably origin of Poltavia fur milch.

Telengit

Area: Altai area of Siberia. Products: Meat and coarse wool.

Ram of the Précoce breed (Omsk Region).

Voloshian
Area: Carpathians to Urals and Caucasus. Products: Meat and coarse wool. Caucasian fat-tailed type and Zackel.

Vyatka
Area: Nolinsk, Kirov, Gorodets, Gorki. Products: Meat and fine white wool. Northern short-tailed coarse woolled and local dense woolled Nolinsk sheep × Précoce and Rambouillet 1936-56. (374,000 in 1969.) Does well in severe conditions as is so often the case where there is Merino blood.

USSR Wool-Meat Breeds

Altai
Area: Siberia. Products: Fine wool and meat. Siberian Merino, Rambouillet or Australian Merino × Mazeav Merino and Caucasus 1936-48. Does well in severe conditions. Very numerous. (5,170,000 in Altai district in 1969.)

Andi
Area: Dagestan. Products: Coarse wool and meat. Good example of original carpet type wool sheep.

Askanian
(Sometimes known as Askanian Rambouillet)
Area: Ukraine. Products: Fine wool and meat. Crossed with Rambouillet and Ukrainian Merino 1925-35. Numerous. (2,115,000 in 1964.)

Azerbaijan Mountain Merino

Area: Azerbaijan. Products: Fine wool and meat. Rambouillet, Caucasus Askaniya and local Merinos 1932–47. (693,000 in 1964.)

Beskaragai Merino
Area: Kazakhstan. Products: Fine wool and meat. Early 1900s in Semipalatinsk in North East Kazakhstan. Mazaev, Novocaucasian Merinos × Kazakh fat-rumped. Later additions of Rambouillet blood in 1932, Askanian in 1934 and Altai in 1947.

Caucasian
(Sometimes called Caucasian Rambouillet)
Area: Northern Caucasus. Products: Fine wool and meat. Novocaucasian Merino improved by American Rambouillet. (6,562,000 in 1969.)

Chaglinsk Merino
Area: Kazakhstan. Products: Fine wool and meat. Local Kazakh breeds crossed with various Merinoes.

Chapan
Area: Kirgizia. Products: Coarse wool and meat. Carpet type wool.

Darvaz
Area: Tajikistan. Products: Coarse carpet wool and meat.

Edilbaev
Area: Western Kazakhstan. Products: Coarse carpet wool and meat. Early developed from fat-tailed local sheep, the Kazakh fat-rumped and Kalmyk. Important carpet wool type sheep. (4,519,000 in 1969.)

Fat-rumped (Type)
Products: Coarse wool and meat. Many

Ram of the meat and fat Edilbaev breed.

varieties exist in the USSR.

Grozny
Area: Dagestan and Kalmyk, ASSR region. Products: Fine wool and meat. From Australian Merino, the Mazeav and the Novocaucasian Merino

Gunib
Area: Dagestan. Products: Coarse black wool, carpet type and meat. Derived from Württemberg Merino with the Dagestan Mountain sheep.

Imeritian
Area: West Georgia. Products: Coarse carpet wool, dairy produce and meat. Caucasian fat-tailed type.

Jaidara
Area: Uzbekistan. Products: Coarse carpet wool and meat.

Kalmyk
Area: Astrakhan. Products: Coarse carpet wool and meat.

Karachaev
Area: North Caucasus. Products: Coarse wool and meat. Caucasian type.

Kazakh Fat-rumped
Area: Kazakhstan. Products: Coarse carpet wool and meat. From Kalmyk, Edilbaev, with Shropshire, Degeres, Saraja and Kargalin. Also found in Sinkiang, China.

Kazakh Finewool
Area: Kazakhstan. Products: Fine wool and meat. Originated in Alma Ata in Kazakhstan from the Précoce and local coarse wool fat-tailed sheep and some Rambouillet blood 1931–45. One of the most important fine-woolled sheep in USSR. (2,925,000 in 1969.)

Kirgiz Fat-rumped
Area: Kirgizia. Products: Coloured

wool and meat. From Précoce Merino, the Alai fat-rumped and fat-rumped Merino.

Kirgiz Finewool
Area: Kirgizia. Products: Fine wool and meat. 1932–56 by interbreeding Württemberg Merino or the Précoce with Rambouillet and Novocaucasian or Siberian Merino × Kirgiz fat-rumped.

Krasnoyarsk
Area: South-central Siberia. Products: Fine wool and meat. Précoce, Rambouillet × Askanian and local fine woolled breeds 1963. (1,580,000 in 1969.)

Orkhon
Area: Mongolia. Products: Medium wool and meat. Altai × (Tsigai × [Précoce × Mongolian]) 1943–61.

Ostrogozhsk
Area: North West Voronezh. Products: Long wool and meat. Romney Marsh with Mikhnov before 1963.

Parkenrskaya
Area: Uzbekistan. Products: Coarse Long wool and meat. Local fat-rumped × Lincoln.

Pechora
Area: Komi. Products: Long wool and meat. Romney Marsh × Russian Northern short-tailed 1937–50.

Salsk Finewool
Area: Rostov, North Caucasus. Products: Fine wool and meat. Rambouillet and Nazaev × Novocaucasian Merino 1930–49.

Saraja
Area: South East Turkmenistan. Products: Coarse wool and meat, also fur

(i.e. skin). Long established derivation from local fat-tailed.

Siberian Rambouillet
Area: Siberia. Products: Probably the same as the Altai.

South Kazakh Merino
Area: Kazalinsk, South Kazakstan. Products: Fine (but uneven) wool and meat.

South Ural
Area: Orenburg, South Urals. Products: Fine wool and meat. Précoce crossed with local coarse wool 1947-68. (192,000 in 1969.)

Soviet Merino
Area: Northern Caucasus, Siberia and the Volga River. Products: Fine wool and meat. Mazaev and Novocaucasian Merino. Also some Rambouillet. Crossed with Siberian Merino 1935-47. Numerous. (8,953,000 in 1969.)

Soviet Rambouillet
Area: Widespread in USSR. Products: Fine wool and meat. From 1920 onwards by Rambouillet × with other local Merinos.

Stavropol Merino
Stavropol region, North Caucasus. Products: Fine wool and meat. Novocaucasian mixed with Rambouillet and Australian Merino 1925-50. (2,840,500 in 1969.)

Sulukol Merino
Area: Kustanai, North West Kazakhstan. Products: Fine wool and meat. Fine wool sheep × Kazakh fat-rumped, Askanian 1958.

Tajik
Area: Tajikistan. Products: Coarse, probably carpet quality wool and meat. Saraja × Hissar with some Lincoln blood

1948-63.

Temir
Products: Coarse carpet quality wool and meat. Variety of Kazakh fat-rumped.

Tian Shan
Area: Kirgizia. Products: Semi fine wool and meat. From Précoce, Novocaucasian Merino and Württemberg Merino 1938-50.

Transbaikal Finewool
Area: Tsitsin, Siberia. Products: Fine wool and meat. From local coarse woolled sheep such as Buryat improved with various Merinos, Altai and Grozny 1927-56.

Turkmen Fat-rumped
Area: Turkmenistan. Products: Coarse wool and meat.

Tushin
Area: Georgia and Armenia, also outside Russia in Turkey. Coarse wool and meat, also dairy produce. Caucasian fat-tailed type.

Uchum
Products: Coarse wool and meat. Larger variety of the Krasnoyarsk.

Volgograd
Area: North East Volgograd. Products: Fine wool and meat. Précoce × Kazakh and Astrakhan fat-rumped, Caucasian and Grozny 1931-63.

USSR Fur (Skin) Breeds

Chushka
Area: Western Ukraine and Moldavia. Products: Fur, dairy produce and meat. Derived from Russian long-tailed and similar to Reshetilovka and Sokolka.

Two Kirgizia shepherds.

Karakul

Area: Uzbekistan, also found in South Africa. Products: Fur and dairy produce. Long established, grown to get lambskins for the Persian Lamb or Astrakhan coat. Lambs are killed second day after birth. (12,547,500 in 1969.)

Karakul Prolific

Area: Ukraine. Product: Fur. Developed in Askania Nova by crossing Karakul with Romanov. Lambing rate as high as 400.

Malich

Area: Crimea. Products: Fur, and dairy. Cross between Karakul and native (i.e. Crimea) coarse woolled sheep.

Poltava Fur-milch

Area: Ukraine. Products: Fur, and dairy. In Ukraine from Russian long-tailed native.

Reshetilovka

Area: Ukraine. Products: Fur and dairy. Derived from Sokolka and Poltava furmilch or Russian long-tailed.

Romanov

Area: Yaroslavl', Kostroma and Ivanova. Products: Fur, also meat and wool. Developed from Russian Northern short-tailed over many years. Important breed 534,000 in 1969. High lambing rates 194/420. Two or four teat udder. Considerable interest shown in this sheep in other countries.

Russian Northern Short-tailed

Area: North European Russia. Products: Fur, also coarse wool. Origin of the Romanov.

Sokolka
Area: Ukraine. Products: Fur, also dairy. From Poltava fur-milch. (164,000 in 1969.)

Asia

Commercially speaking, the importance of these sheep is mainly for carpet wool. This wool can be extremely varied from fine to very coarse, dirty, liable to cause the anthrax and generally regarded with distaste by the rest of the wool trade. There is also another interesting side of fibre production in this area, namely the rarer hairs such as camel, widely used for ladies' coatings, and dressing gowns, although many so-called camel cloths may have only 5% and that camel hair waste, the rest being wool.

Asia also possesses specimens of the original wild sheep—the Mouflon and the Urial. The first looks like an antelope and has a wide range over the Near East and Central Asia. The other is more confined, being found mainly in the Elburz Mountains. Both have inner and outer coats.

As a source of reference Mason's lists are, as always, invaluable. For China there is Epstein's *Domestic Animals of China.*

In the following lists and brief descriptions of Asian breeds a geographical division is employed: (a) India and Pakistan, (b) Central Asia and (c) China.

(a) Breeds from India and Pakistan

According to the figures given in *British*

Sheep, India is not of outstanding importance in the world of sheep, the total number there being about 43 million, which can be compared with the 70 million in New Zealand, and the 39 million in Turkey, to make a closer geographical comparison. The main product is coarse wool of the mixed type suitable for carpets and known to the wool trade as East Indian. It is the main export and the meat and milk are almost entirely consumed locally. Pakistan has approximately 17½ million sheep of a similar type.

Balkhi
Area: North West Afghanistan and North West Frontier Province, Pakistan. Products: Coarse carpet wool and dairy. Caucasian flat-tailed type.

Baluchi (Type)

Area: Baluchistan, East Persia and Sind, Pakistan, also East Afghanistan and East Iran. Products: Meat and dairy. Widely spread and several specific breeds.

Bellary
Area: East Mysore and West Andhra Pradesh, India. Products: Coarse carpet wool and meat.

Bhadarwah
Area: Jammu and North East Punjab, India. Products: Coarse wool, also for pack carrying.

Bhakarwal
Area: South West Kashmir, India. Products: Coarse carpet wool and meat. The word 'Bhakarwal' means 'class of nomads'.

Biangi
Area: Tibet and North East Punjab,

Newborn Karakul lambs.

Bibrik
Area: North East Baluchistan. Products: Coarse carpet wool and meat.

Bikaneri
Area: Rajasthan, India. Products: Coarse carpet wool. One of the most important of Indian breeds.

Chanothar
Area: South Rajasthan and North Gujarat, India. Products: Coarse carpet wool, and dairy.

Damari
Area: South of North West Frontier Province, Pakistan. Products: Dairy and wool.

Deccani
Area: Maharashtra, India. Products: Coarse carpet wool and meat. One of the better known Indian carpet wool producing breeds.

Dumari
Area: Baluchistan, India. Products: Coarse carpet wool and meat. Same as the Harnai.

Gujarati
Area: Gujarat, India. Products: Coarse carpet wool and dairy.

Gurez
Area: North Kashmir, India. Products: Coarse wool and dairy.

Hairy (Type)
Products: Coarse, usually carpet wool. Name given to widely spread sheep all having some similarity to the wild sheep. Many varieties.

Harnai
Area: Quetta, Baluchistan, Pakistan. *See* Damari.

Hashtnagri
Area: North West Frontier Province, Pakistan, also Afghanistan. Products: Coarse carpet wool, also meat and dairy.

Hassan
Area: Mysore, India. Products: Coarse carpet wool. Probably originated from Deccani crossed with Nellore.

Hissar Dale
Area: East Punjab, India. Products: Short fine wool. From Australian Merino ($\frac{7}{8}$) × Bikaneri 1920s. Said to be only one flock. South Indian hairy type.

Jaffna
Area: Sri Lanka. Product: Meat. South Indian hairy type.

Jalauni
Area: South West Uttar Pradesh, India. Products: Coarse carpet wool and meat.

Kaghani
Area: Hazara, North East Peshawar, Pakistan. Products: Dairy, also coarse wool.

Karnah
Area: Muzaffarabad, North West Kashmir, Pakistan. Product: Short wool.

Kathiawari
Area: North Bombay. *See* Gujarati.

Khurasani
Area: East Iran and Baluchistan. Products: Meat, coarse wool and dairy.

Kuka
Area: Tharparkar, Sind, Pakistan. Products: Coarse carpet wool and dairy.

Lohi
Area: South Punjab, Pakistan. Products: Coarse wool, meat and dairy.

Mandya
Area: South West Mysore, India. Pro-

duct: Meat. South Indian hairy type.

Nellore
Area: South East Andhra Pradesh, India.
Product: Meat. Important breed of the
South India hairy type.

Rakhshani
Area: West Baluchistan, Pakistan. Products: Coarse wool, meat and dairy.

Short Tailed (Type)
Represented in area by many breeds.

Sikkim
Area: India. Products: Meat and coarse
wool.

South India Hairy (Type)
Product: Meat. Many variations in area.

Thal
Area: Thal Desert, West Punjab, Pakistan. Products: Coarse carpet wool and
meat.

Tirahi
Area: Tirah and Kurram, North West
Pakistan. Products: Coarse wool and
meat.

Vicanere *See* Bikaneri
Some of the best known carpet wools
from India come to the market under the
name Vicanere.

Waziri
Area: Waziristan, North West Pakistan.
Products: Coarse wool and meat.

(b) Breeds from Central Asia

An important sheep producing area: In
terms of sheep numbers the leading
countries are:

Afghanistan	23½	million
Iran	37	million
Iraq	19	million
Turkey	39	million

(From *British Sheep*)

Most of the sheep produce coarse
wool usually of the carpet, i.e. mixed
type and this is the main commercial
product. But the meat and dairy products are used locally.

Amasya Herik
Area: North Anatolia, Turkey. Pro-
ducts: Coarse wool, meat and dairy.
Similar to Dǎgliç.

Arabi
Area: Israel, South West Iran, South
Iraq and North East Arabia. Products:
Meat, also coarse wool. Near East fat-
tailed type.

Argali
Area: Central Asia, Himalayas to Mongolia. One of the four important primitive sheep.

Awassi
Area: Israel, South West Iran, South
Iraq and North East Arabia. Products:
Meat, also dairy and coarse wool. Near
East fat-tailed type and the typical sheep
of that area.

Awassi Ram, Israel.

India. Products: Coarse carpet wool, also for pack carrying.

Dăğliç
Area: West Anatolia. Products: Coarse carpet wool and dairy. One of the best known sheep of the area.

Farahani
Area: West Central Iran. Product: Coarse carpet wool.

Hejazi
Area: Arabia. Product: Meat. Fat-rumped type.

Iran Fat-tailed (Type)
Specialised type of Middle East fat-tailed.

Iraq Kurdi
Area: North East Iraq. Products: Meat, coarse wool and dairy.

Kalaku
Area: North East Iran. Product: Coarse wool. Iran fat-tailed type.

Kamakuyruk
Area: North West Anatolia. Products: Coarse wool, meat and dairy. Probably Dağliç × Kivircik.

Kandahari
Area: South West Afghanistan. Product: Coarse carpet wool.

Karayaka
Area: North Antolia, Turkey. Products: Coarse wool, meat and dairy.

Kermani
Area: South East Iran. Products: Meat, coarse wool and dairy.

Khurasani
Area: East Iran and Baluchistan. Products: Meat, coarse wool and dairy.

Kivircik
Area: North West Turkey, also Greece. Products: Meat, dairy and coarse wool. Similar to Karnobat of Bulgaria.

Kizil-Karaman (Red Karaman)
Area: East Turkey. Products: Coarse carpet wool, meat and dairy. Middle East fat-tailed type.

Kurdi (Type)
Area: Kurdistan and North East Iran. Product: Coarse carpet wool.

Near East Fat-tailed (Type)
Area: Also found in North East Africa. Products: Coarse carpet wool, meat and dairy. Perhaps most important breed in Central Asia. Many varieties.

Nejdi
Area: Central Arabia. Product: Coarse carpet wool.

Pirlak   See Kamakuyruk.

Red Karaman
Area: East Turkey. Products: Coarse wool, meat and dairy. Important sheep of Middle East fat-tailed type.

Tuj
Area: Çildir, North Eastern Turkey. Products: Meat and coarse wool.

Turkish Merino
Area: North East Anatolia. See Merino section.

Urial or Ovis Vignei
Area: Central Asia—variations in Kashmir, Afghanistan, North Baluchistan and South Russian Turkestan. Most of our domestic sheep today come from ancestors of this wild sheep.

White Karaman
Area: Central Anatolia, Turkey. Products: Meat, dairy and coarse wool.

There is a good account of this sheep in Mason: *Sheep Breeds of the Mediterranean*.

(c) Breeds from China

There are reported to be seventy-one million sheep in this vast area and the plan is to increase it to 167 million. At present 83% of the wool is carpet quality, 15% crossbred and 2% Merino. Pelt and food production is also important. Comparatively little of this material finds its way onto the world markets and it is in fact the rare goat fibre, Cashmere, that is China's most important contribution to the world fibre market. As in other similar areas, the sheep is also produced for meat and dairy purposes but neither of these commodities play any part in international trade. H. Epstein *Domestic Animals of China* is the main source of information.

Chowpei
Area: Hupeh. Products: Coarse carpet wool. New breed of fat-tailed type.

Han-yang
Area: Hilly areas in Honan and in parts of Shansi, Hopeh, Shantung and Kiangsu. Products: Semi fine wool and meat. Lambs wool (Shantung lambs) said to be 64s (22 microns).

Hei or Ningsia Black Sheep
Area: Ningsia. Products: Coarse carpet wool-pelt (or fur). Probably from Karakul rams with Mongolian fat-tailed ewes.

Hetian-yang
(Or Hotien-yang. Also, according to Mason, Khotan.) Area: West Sinkiang. Products: Coarse carpet wool, meat and pelt.

Hu-yang
Area: South Kiangsu. Products: Coarse carpet wool and pelt. Probably has highest lambing rates of any sheep. Litters of 5 to 6 lambs are known.

Kuche
Area: West Sinkiang. Products: Pelt, also coarse wool.

Luan
Area: Shansi. Products: Coarse carpet wool.

Mongolian
(Or Chinese fat-tailed.) Products: Coarse carpet wool, meat and dairy. Most important of Chinese sheep and spreading. Many varieties.

Native and Crossbred
Area: North West China. Products: Semi fine wool and meat. Many native sheep of the traditional type are being improved with Merino blood.

North East China Merino
Products: Fine wool and meat. Attempts are being made to improve the old native breeds with Merino blood, usually in the form of Sinkiang fine-woolled breeds.

North West China Merino
Products: Fine wool and meat. New breed evolved from Mongolian fat-tailed and Tibetan thin-tailed ewes with Sinkiang fine-woolled, Caucasian Rambouillet and Stavropol Merino rams.

Showyang
Area: Shansi. Products: Coarse carpet wool.

Sinkiang finewool
Area: Sinkiang. Products: Fine wool and meat. In 1935 under guidance of Soviet zoo-technicians invited by a local

war lord, the breed was produced by cross breeding Kazakh fat-rumped and Mongolian with Novocaucasian Merino and Précoce.

Taiku

Area: Shansi, Hopel and Shensi. Products: Coarse carpet wool.

Tan-yang

Area: North Ningsian and neighbouring areas of Kansu, Shensi and Inner Mongolia. Products: Pelt, coarse wool and meat. Lambs are killed at 16–30 days.

A Tan Yang ram.

Tibetan

Area: Tibet, Szechwan, Tsinghai, Yunnan, Kweichow and South Kansu. Product: Coarse carpet wool. Important breed with several variations. Sheep very hardy and the wool is reckoned the finest carpet type produced in China.

Tung-yang

Area: Lo Ho Valley, Shensi. Products: Medium wool and meat. Said to have been introduced to China by Mohammedan immigrants during the T'ang dynasty (A.D. 618–906), and one of the longest established Chinese breeds.

Africa

Africa is the most difficult to describe of all sheep producing areas. First as regards numbers, there were said to be 137 million in 1973 (*British Sheep*, Production Year Book [1973]) and by far the most important area is South Africa with 30·5 million in 1973 (*Wool Facts*, IWS [1978]) of which about 80% were Merinos. South Africa comes second only to Australia as a great Merino producing country. This area has been described in the section on the Merino sheep. There are also a few other Merinos, notably in Kenya, but basically other African sheep are derived from the Fat-tailed type but the wool is not now particularly suitable for carpets and in this respect does not compare with the Asiatic and Indian versions of that breed. The commercial importance of this coarse wool is therefore comparatively small and the main reason for keeping sheep is the local meat and dairy trade. One or two exceptions deserve mention, there are many Karakuls in what was South-West Africa, and Mohair is also produced in South Africa.

Once again my main source of information for non-Merino and British breeds has been Mason *A Dictionary of Domesticated Breeds* and in some cases his *Sheep Breeds of the Mediterranean*. The I.W.S. and individual authorities in South Africa have been most helpful with information regarding such breeds as the Karakul and the Dorper. During my years in the wool trade I knew South African and Kenyan Merino well and also valued and occasionally used other non-merino types from these two areas.

Herdsmen sell wool to commercial department in Tibet.

40 White Faced Dartmoor.

41 Jacob.

42 Icelandic.

43 Friesian.

African Breeds

Abyssinian
Area: Ethiopia. Products: Meat and dairy (milk rather than cheese). East African fat-tailed type.

African long fat-tailed (Type)
Many individual breeds.

African long-legged (Type)
Many individual breeds. This type would seem to be derived from the sheep known in ancient Egypt.

Africander
Area: South Africa. African long-legged type.

Algerian Arab
Products: Meat and coarse wool. Several sub-varieties.

Atlantic Coast (Type)
Area: Morocco. Several breeds.

Ausimi
Area: Lower Egypt. Products: Coarse wool and meat. Most numerous breed in Egypt.

Barbary (Type)
Area: North Africa. Products: Dairy, coarse wool and meat. Middle East fat-tailed type.

Barki
Area: North West Egypt. Products: Coarse wool, meat and dairy. Derived from Barbary type.

Beni Ahsen
Area: North West Morocco. Products: Meat and coarse wool.

Beni Guil
Area: East Morocco and West Algeria. Products: Meat and coarse wool.

Berber
Area: Morocco. Products: Meat and coarse wool. Ancient breed found in Mountains of Morocco.

Blackhead Persion
Area: South Africa. Product: Meat. From 1868 onwards derived from Somali.

Cape Native
Area: South Africa. Products: Wool and meat. Although not apparently a recognised breed today, much South African wool from such sheep as Namaqua was sold under this name. The wool was surprisingly fine but rather spoilt by kemp.

Congo long-legged
Area: Kibali-Ituri, North East Congo. African long-legged type.

Dorper
Area: South Africa. Product: Meat. From 1940 onward at Grootfontein from Dorset Horn × Blackheaded Persian.

Doukkala
Area: South West Morocco. Products: Coarse wool and meat. Atlantic coast type.

Dwarf (Type)
Several varieties.

East African fat-tailed (Type)
Important African type. Many varieties.

Fulani
Area: Senegal to Cameroun. Product: Meat. West African long-legged type.

Ghimi
Area: Fezzan, Libya. Products: Coarse wool and meat. Derived from Algerian Arab × Libyan Barbary.

Hottentot

Area: South Africa. Now extinct but at one time important. Derived from Near East fat-tailed and Ancient Egypt long-tailed. Origin of present Africander.

Karakul

Area: South Africa. Product: Fur. Essentially for lambskins yielding Astrakhan fur. Numerous in USSR.

Karakul Sheep, South Africa.

Libyan Barbary

Area: Libya. Products: Meat, coarse wool and dairy.

Macina

Area: Central delta of Niger, Mali, West Africa. Products: Coarse wool.

Madagascar

Area: Madagascar. Products: Meat.

Maghreb (Type)

Typical sheep of Morocco and Tunis. Several varieties.

Malawi

Area: Malawi. African Long fat-tailed.

Masai

(Sometimes called Tanganyika short-tailed.) Area: Tanzania, Kenya and Uganda. Product: Meat. East African fat-tailed type.

Maure

Area: North of West Africa. Product: Meat. West African long-legged type.

Namaqua Africander

Area: North West of Cape Province and south of South West Africa. Product: Wool. Variety of Africander, originally from Namaqua variety of Hottentot.

Nguni

South East Africa. African long fat-tailed type.

Northern Sudanese

Sudan, north of 12°N., and Western Eritrea. Products: Dairy and meat. African long-legged type.

Rahmani

Area: Beheira, Lower Egypt. Products: Coarse wool and meat.

Rhodesian

Area: Rhodesia. African long fat-tailed type.

Ronderib Africander

Area: North Central Cape Province, South Africa. Product: Meat. Originally derived from Cape variety of the Hottentot.

Saidi

Area: Egypt. Products: Coarse wool and meat.

Sanabawi

Area: Egypt. Products: Coarse wool and meat. Similar to Saidi.

Sidi Tabet Cross

Area: Tunisia. Product: Medium black wool. New development from Portuguese black Merino × Thibar.

44 North Ronaldsay.

45 Champion Romney New Zealand Ram.

46 Sheep in the yards, Canterbury, New Zealand.

47 Cretan shepherd and flock.

Somali
Area: Somalia, also East Ethiopia and North Kenya. Product: Meat. Origin of blackheaded Persian.

Sudanese (Type)
Area: Sudan. Product: Coarse wool. Several varieties.

Tadla
Area: Morocco. Products: Meat and coarse wool. Main breed of western plateau of Morocco.

Tadmit
Area: Algeria and Tunisia. Products: Meat and medium wool. From Algerian Arab c.1925 perhaps with some Merino.

Tanganyika long-tailed
Area: Tanzania. Product: Meat.

Thibar
Area: Tunisia. Products: Meat and medium wool. The breed was created c.1911 by crossing Tadmit with Arles Merino.

Tswana
Area: Botswana and South West Rhodesia. African long fat-tailed type.

Tuareg
Area: Sahara. Product: Meat.

Tunisian Barbary
Area: Tunisia. Products: Meat and coarse wool. Origin of Companian Barbary and the Sicilian Barbary.

West African Dwarf
Area: South of West Africa.

West African long-legged (Type)
Hairy type, several varieties.

White Dorper
Area: Transvaal, South Africa. Product: Meat. From 1946 on by crossing Dorset

Horn with the Blackheaded Persian in Pretoria.

Zemmour
Area: North West Morocco. Products: Meat and coarse wool. Atlantic coast type.

Europe

European breeds present a problem which is best considered by looking at them firstly from the Merino–British breed base already emphasised, and secondly from a topographical or geographical viewpoint. The Merino is definitely more important than the British. The scale of Merino importations from Spain has already been emphasised and naturally enough, they have left their mark, which range from the basic traces still existing in Spain to important developments later in Germany and France. The British breeds have had less effect, far less than they have had in other parts of the world. Concerning the topographical element, Europe can be divided, in sheep breeding as in so many other ways, into the industrial north, west and centre, and the agricultural south, i.e. the more backward (industrially speaking) periphery. Naturally there are more sheep in the latter although two of the more important European breeds of the second group could well be classed as rare and a European Rare Breed Society is badly needed.

In the following lists and brief descriptions, a geographical division is employed:

	Million
(a) Spain and Portugal	19·500
(b) France	10·218
(c) Italy	7·770

(d) East Europe   *Million*
     Albania   1·163
     Bulgaria   9·921
     German
       Democratic
       Republic   1·657
     Hungary   2·259
     Poland   3·050
     Rumania   14·455   32·505
(e) Miscellaneous:
     German
       Federal
       Republic   ·908
     Greece   7·620
     Norway   1·648
     Yugoslavia   7·776   17·952

(Also from Switzerland, Belgium, Austria, Finland, Iceland and Sweden.)

European sheep are produced almost entirely for home consumption—meat, dairy and wool. In so far as there are other purposes, those for export are for breeding, and for cheese production, for example Roquefort.

Sources here as elsewhere in this section are Mason's *A Dictionary of Livestock Breeds* and where applicable the same author's *Sheep Breeds of the Mediterranean* with its excellent descriptions and photographs. I have had much assistance from many articles published in agricultural and other papers, notably those by Dr M. L. Ryder with whom I have also had several valuable discussions. Several of the European breed societies have sent me helpful pamphlets and given much useful information. Finally, I have when possible seen some of the breeds but the only wool from Europe that I have ever used was that of the Arles Merino.

(a) Spanish and Portuguese Breeds

The sheep of the Iberian peninsular can be divided into two groups, those that are derived from Spain's great contribution to sheep breeding, namely the Merino, and those derived from the 'Churro' which is Spain's word for coarse wool.

Algarve Churro
Area: West Spain from Zamora to Meat and coarse wool. From Spanish Churro c.1870–90 imported from Andalusia.

Andalusian Churro
Area: West Spain from Zamore to Andalusia. Products: Dairy, meat and coarse wool. Largest variety of Spanish Churro.

Andalusian Merino
Area: West Andalusia. Products: Meat and semi fine wool.

Aragon
Area: Aragon, also Navarre, Catalonia and Castellón. Products: Meat and medium wool. Rasa type, i.e. smooth fleece as opposed to rough, open-fleeced Churro.

Badano
Area: Terra Quente, North East Portugal. Products: Meat, coarse wool and dairy. Portuguese Churro type.

Beira Baixa
Area: Castelo Branco, Portugal. Products: Dairy and fine wool. Milk producing Merino strain (250,000).

Black Merino
Area: Portugal. *See* Merino section.

48 Gathering sheep in Wales.

49 Inspecting teeth at sale of Welsh Mountain Rams.

50   Dipping, UK.

Bordaleiro (Type)
Area: Portugal. Products: Dairy, meat and medium wool. From Merino × Churro and now covers all kinds of these intermediate sheep.

Bragança Galician
Area: Terra Fria, North East Portugal. Products: Meat and coarse wool. Portuguese Churro type.

Campaniça
Area: South Beja, South Portugal. Products: Meat, medium wool and dairy. Bordaleiro type.

Castilian
Area: Valladolid and surrounding areas. Products: Meat, dairy and wool. Raso type.

Churro (Type)
Area: North West Spain, also found in Cantabria and the Pyrenees, and in Northern Portugal. Products: Coarse wool and dairy. Many sub-groups including the Portuguese Churro and Spanish Churro.

Churro do Campo
Area: North Castelo Branco, Beira Baixa, Portugal. Products: Meat and coarse wool. Portuguese Churro type.

Entrefino (Type)
Area: Spain. Products: Meat and medium wool. Originated from Churro × Merino many specialised types.

Entre Minho e Douro
Area: North West Portugal. Products: Medium wool and meat. Bordaleiro type.

Fonte-Bôa Merino
Area: Portugal. Products: Fine wool and meat. From Rambouillet × Spanish Merino at the Fonte-Bôa Zootechnical Station 1902-26.

Lacho
Area: Vascongadas and Navarre, Spain. Products: Dairy and coarse wool.

Mancha
Area: La Mancha, New Castile. Products: Dairy, wool and meat. Entrefino type.

Miranda Galician
Area: Terra Fria, North East Portugal. Products: Meat and coarse wool. Portuguese Churro type.

Mondego
Area: West Guarda, Beira Alta, Portugal. Products: Meat, dairy and coarse wool. Portuguese Churro type.

Portuguese Churro (Type)
Area: North East Portugal. Many specialised breeds.

Portuguese Merino   *See* Merino section.

Saloia
Area: Lisbon, Portugal. Products: Dairy, medium wool and meat. Bordaleiro type.

Segura
Area: Murcia, Spain. Products: Medium wool. Entrefino type.

Serra da Estrêla
Area: North Central Portugal. Products: Dairy, medium wool and meat. Bordaleiro type.

Spanish Churro
Area: North and West Spain. Products: Dairy, meat and medium wool.

Talavera
Area: West Toledo. Products: Dairy, meat and medium wool.

(b) French Breeds

France has ten million sheep. Her great contribution to sheep breeds is in the Rambouillet Merino (see Merino section) and in the dairy breeds used for cheese production.

Aure-Campan
Area: South East Hautes-Pyrénées. Products: Meat and medium wool. Migratory breed.

Avranchin
Area: South Manche, North France. Products: Meat and long wool. Similar to Cotentin, originated from Leicester (+Southdown and Romney) × local sheep 1830–1900.

Basque
Area: South West France. Products: Dairy, meat and coarse wool. Origin of Basque-Bearn and now regarded as a variety of that breed.

Basque-Béarn
Area: South East Basses-Pyrénées. Products: Dairy, meat and coarse wool. Since 1965 from Basque + Béarn.

Béarn
Area: South West France. Products: Dairy, meat and coarse wool. West Pyrenean type; origin of and now variety of Basque-Béarn.

Berrichon
Area: Central France. Products: Meat and short wool.

Bizet
Area: Haute Loire, Cantal. Products: Meat and wool.

Blanc du Massif Central
Area: Lozère, Ardèche, Gard and Hérault. Products: Meat, medium wool and dairy. From Causse × Lacaune, early twentieth century. Given its present name in 1965.

Bluefaced Maine (Type)
Area: North West France. Products: Meat and long wool. From Leicester (and Wensleydale) imported 1855–80 and crossed with local breeds.

Boulonnais
Area: North France. Products: Meat and long wool. From Dishley Leicester in nineteenth century with some Merino blood and also crossed with Artois.

Causses (Type)
Area: South France. Product: Coarse wool. Many specified breeds in southern France.

Central Plateau (Type)
Area: Massif Central. Several specified types.

Central Pyrenean
Area: Central and East Pyrénées. Products: Meat and medium wool. From original Pyrenean type improved by Spanish Merino.

Charmoise
Area: West Central France. Products: Meat and short wool. From Romney × 1944–52 (Berrichon-Merino-Sologne-Touraine). Originated at La Charmoise farm, Loir-et-Cher.

Cher Berrichon
Area: Berry. Products: Meat and medium wool. Long established, from Merino in eighteenth century, Southdown, Leicester and Dishley Merino in nineteenth century, also crossed with Champagne and Boischaut varieties of Berrichon.

51   Scottish Agricultural show—sheep in the ring.

52   Penderyn sheep sales, Wales.

Corsica

Area: Corsica. Products: Dairy, meat and coarse wool.

Cotentin

Area: North Manche. Products: Meat and long wool. Nineteenth century from Leicester crossed with local breed (Berca). Similar to Avranchin.

Dishley Merino (or Ile de France)

Products: Meat and medium wool. Nineteenth century from Leicester × Merino, renamed Ile de France 1922.

French Alpine

Area: Hautes Alpes. Products: Meat and medium wool. From local with Prealpes du Sud blood.

French Blackheaded

Area: Moselle. Products: Meat and short wool. From Suffolk, Hampshire Down, Oxford Down, Southdown and since 1945 German Blackheaded mutton.

Ile de France   See Dishley Merino.

Indre Berrichon

Area: Berry. Products: Meat and short wool. From Champagne and Crevant varieties of Berrichon.

Lacaune

Area: Tarn, Southern France. Products: Dairy, meat and medium wool. Important breed of the Roquefort type. Probably originated from Merino and Barbary crossed by Caucasus.

Landes

Area: South West France. Products: Meat and coarse wool. Disappearing gradually.

Limousin

Area: Corrèze. Products: Meat, wool and dairy. Central Platau type.

Lot Causses

Area: Lot. Products: Meat and coarse wool. Causses type.

Lourdes

Area: South West Hautes-Pyrénées. Products: Meat and medium wool. West Pyrenean type.

Manech

Area: South West Basque country. Products: Dairy, meat and coarse wool. Similar to Spanish Lacho.

Mouflon

Area: Corsica and Sardinia. One of the four main primitive types of sheep that still exist.

Préalpes du Sud

Area: Drôme, South West Hautes-Alpes and North East Vaucluse. Products: Meat and short wool. Mason: *Sheep Breeds of the Mediterranean* has a good account of this sheep.

Précoce

See Merino section.

Pyrenean (Type)

Leading varieties are Manech, West Pyrenean and Central Pyrenean types.

Roquefort Breeds (Type)

Area: South France. Product: Dairy. One of the most important of French breeds as its milk is used for Roquefort cheese. Includes Larzac, the original type.

Sologne

Area: East Loir-et-Cher, Central France. Products: Meat and short wool. Similar to Berrichon.

Tarascon

Area: South France. Products: Meat and medium wool. Central Pyrenean type.

Thônes-Marthod
Area: Savoy. Products: Meat and coarse wool.

Velay Black
Area: Haute-Loire. Products: Meat and medium wool. Central Plateau type.

(c) Italian Breeds

Italy has approximately 8 million sheep, almost all bred for local use and without any major part in the world picture. As elsewhere a division can be made, those with Merino blood, those with British blood (comparatively few) and native breeds which can be subdivided into mainland and island, of which the Sardinian sheep are both the most numerous and most interesting in the country. The three leading breeds are:

	Million
Sardinian	2·340
Sopravissana	1·050
Apulian Merino	·950

Alpago
Area: Belluno, Venetia. Products: Medium wool, meat and dairy. Lop-eared Alpine type.

Altamura
Area: Bari, Apulia. Products: Dairy, meat and coarse wool. Moscia type.

Apennine (Type)
Area: Pennines from Emilia to Abruzzi. Products: Dairy, meat and medium wool. From Merino crossed with local breeds such as Bergamo and Sopravissana. Many specific breeds.

Apulian Merino See Merino section

Bergamo
Area: Lombardy. Products: Meat and coarse wool. Basic surviving form of

Lop-eared Alpine type.

Biella
Area: Piedmont. Products: Meat and coarse wool. Lop-eared Alpine type.

Briga
Area: Là Brigue, borders of South East France and North West Italy. Products: Dairy, meat and coarse wool. Similar to Frabosa. Decreasing in numbers.

Calabrian
Area: Calabria. Products: Coarse wool and dairy. Moscia type.

Campanian Barbary
Area: Campania. Products: Dairy, meat and coarse to medium wool. From Tunisian Barbary possibly crossed with thin tailed breed.

Carapelle
Area: Foggia, Apulia. Products: Medium black wool, dairy and meat. Nearly extinct.

Cascia
Area: Umbria. Products: Dairy, meat and medium wool. Local variety of Apennine.

Casentino
Area: Arezzo, Tuscany. Products: Dairy, meat and medium wool. Apennine type with Sopravissana blood.

Chiana
Area: Tuscany. Products: Dairy, meat and medium wool. Apennine type with considerable Sopravissana blood.

Chieti
Area: Abruzzi. Products: Dairy, meat and medium wool. Apennine type.

Comiso
Area: South East Sicily. Products: Dairy, coarse wool and meat. From Maltese

crossed with Sicilian in late nineteenth century and early twentieth.

Corniglio
Area: Emilia. Products: Dairy, meat and medium wool. Local variation of Apennine with Merino blood.

Frabosa
Area: Ligurian Alps, Cuneo. Products: Dairy, meat and coarse wool. Similar to Langhe.

Friuli
Area: Udine, Venetia. Products: Meat, medium coarse wool and dairy. Lop-eared Alpine type, typical three-purpose family sheep.

Garessio
Area: Ligurian Alps. Products: Dairy, meat and coarse wool. Local breed of Apennine type.

Garfagnana
Area: North West Tuscany. Products: Dairy and meat. Apennine type similar to Massa.

Improved Apulian
(Or Apulian Merino.) Area: South Italy. *See* Merino Section.

Lamon
Area: Belluno, Venetia. Products: Coarse wool and meat. Lop-eared Alpine type.

Langhe
Area: Piedmont. Products: Dairy, meat and coarse wool. Lop-eared Alpine type.

Lecce
Area: Apulia. Products: Dairy, coarse wool and meat. Moscia type.

Massa
Area: Versilia, North West Tuscany. Products: Dairy, meat and coarse wool.

Alpine type similar to Garfagnana.

Moscia (Type)
Area: South Italy. Products: Dairy, coarse wool and meat. Probably connected with Zackel type.

Paduan
Area: Venetia. Products: Dairy, meat and medium wool. Lop-eared Alpine type.

Pagliarola
Area: Abruzzi and Molise. Products: Coarse/medium wool, meat and dairy. Apennine type, the name means 'straw eater'.

Perugian Lowland
Area: Umbria. Products: Dairy, meat and coarse wool. Apennine type with Bergamo blood.

Sambucco
Area: Stura and Maira valleys, Ligurian Alps. Product: Medium wool. Similar to Garessio.

Sardinian
Area: Sardinia. Products: Dairy, meat and coarse wool. Most numerous of Italian sheep. Similar to sheep illustrated in early island art c.2000 B.C. Anybody who has ever visited Sardinia will have noticed this breed.

Savoy
Areas: West Turin, Piedmont. Products: Dairy, meat and coarse wool.

Sicilian
Area: Sicily. Products: Dairy, meat and coarse wool. There is a good account in *Sheep Breeds of the Mediterranean*. Wool used for carpets and stuffing mattresses, milk now most important product, mainly for cheese. Probably just about what the old pre-Roman sheep pro-

duced in Europe.

Sicilian Barbary
Area: Central Sicily. Products: Dairy, meat and coarse wool. From Tunisian Barbary × Sicilian.

Siena
Area: Arezzo, Tuscany. Products: Dairy, meat and coarse wool. Apennine type.

Sopravissana
Area: Central Apennines, especially Latium and Umbria. Products: Semi fine wool, dairy and meat. Important Italian breed, from Visso crossed with Spanish Merino and Rambouillet in eighteenth and early nineteenth centuries and improved by American and Australian Merino in the twentieth century.

Taro
Area: Emilia. Products: Dairy, meat and coarse wool. Apennine type.

Upper Visso See Sopravissana.

Varese
Area: Lombardy. Products: Meat and coarse wool. Lop-eared Alpine type, similar to, or variety of the Bergamo.

Varzi
Area: Emilia. Products: Dairy, meat and coarse wool. Apennine type.

Vicenza
Area: North East Vicenza. Products: Coarse wool and meat. Local variety of the Lamon.

Visso
Area: Central Apennines. Products: Dairy, meat and coarse wool. Apennine type.

(d) Eastern European Breeds (i.e. from Albania, Bulgaria, German Democratic Republic, Hungary, Poland and Rumania)

As far as sheep are concerned, the important areas are Bulgaria and Rumania which has 24 million of the total of 32 million. The products are almost everywhere used locally, so play no part in world markets.

Common Albanian
Area: South Albania. Products: Dairy and coarse wool. Small sheep from the plans, of the Zackel type.

Danube Merino See Merino section.

Fagas See Pomeranian.

Hungarian Combing Wool Merino
Products: Fine wool and dairy. From Racka (+ Bergamo and Württemberg) improved by Rambouillet.

Hungarian Mutton Merino
Products: Meat and fine wool. From Hungarian combing wool Merino improved by Rambouillet, also Précoce and German mutton Merino.

Karnobat
Area: South East Bulgaria. Products: Dairy and semi fine, usually black wool. From black Tsigai.

Lowicz
Area: Lodz, Poland. Products: Long wool and meat. Between 1924-39 bred from Romney Marsh × local (white Swiniarka with Merino blood).

Luma
Area: Albania. Products: Meat, coarse wool and dairy. Largest of Albania

Flock of Racka on the Bugac Plain in Eastern Hungary.

breeds. Sometimes called the Ruda.

Palas Merino   See Merino Section.

Pleven Blackhead
Area: North Bulgaria. Products: Dairy
and coarse wool. From Bulgarian native
improved by Tsigai.

Polish Heath
Area: Bialystok, Poland. Products:
Coarse wool and fur (pelt). Being im-
proved by the Romanov, USSR.

Polish Longwool (Type)
Area: Poland. Product: Long wool.
Originally from East Friesian, Leine,
Texel and especially Romney Marsh ×
local (i.e. Swiniarka). Several varieties.

Polish Merino   See Merino section.

Polish Zackel
Area: South Poland. Products: Coarse

wool and dairy, also for fur (pelt).

Pomeranian
Area: Pomorze, Poland. Products:
Dairy, long wool and meat. Originally
considered a marsh type but now a longwool. From Friesian in eighteenth century with East Friesian and Wilster-
marsh blood between 1929 and 1939.
Later Texel blood was added.

Racka
Area: Hungary. Products: Dairy-meat,
and coarse wool. Zackel type.

Rila Monastery
Area: Bulgaria. Products: Dairy and
coarse wool. Local Bulgarian Native
improved by Tsigai.

Shkodra
Area: North West Albania. Products:
Coarse wool, dairy and meat. Variation

of the Albanian Zackel type.

Shumen
Area: North East Bulgaria. Product: Coarse coloured wool. Originally probably from Tsigai.

Skudde
Area: East Prussia. Product: Coarse wool. Similar to Pomeranian, now extinct.

Šumava
Area: Bohemia, Czechoslovakia. Product: Coarse wool. Zackel type, similar to Valachian.

Svishtov
Area: North Bulgaria. Product: Coarse wool. From Bulgarian Native improved by other blood.

Swiniarka
Area: Poland. Product: Coarse wool and dairy. From (originally) Romney Marsh.

Turcana
Area: Rumania. Products: Dairy and coarse wool. Zackel type.

Valachian
Area: East Moravia and Slovakia. Products: Dairy, meat and coarse wool. Zackel type.

White Klementina
Area Bulgaria. Products: Coarse wool and dairy. From White Karnobat and Romanian Tsigai with some Merino blood 1910, at State Farm Klementina (now G. Dimitrov) near Plovdiv.

White South Bulgarian
Area: Bulgaria. Products: Coarse wool, dairy and meat. Local Bulgarian Native sheep improved by Tsigai.

(e)   Miscellaneous Breeds

The remaining areas of Europe could be further sub-divided into two sections: firstly Yugoslavia and Greece, each with 8 million sheep, have by far the most, and secondly the more industrial states which have only 5 million, some of which are, however, of considerable commercial importance in the breeding field. With the first group the principal product is milk, mainly for cheese. The island sheep are particularly interesting often highly productive when kept in very small flocks around the home.

Bardoka
Area: Metohija, South West Serbia, Yugoslavia. Products: Dairy, meat and coarse wool. Pramenka type.

Black-Brown Mountain
Area: Switzerland. Products: Short wool and meat. The breed was produced by combining Frutigen, Jura, Roux-de-Bagnes, Saanen and Simmental 1938.

Bosnian Mountain
Area: Central and West Bosnia and Hercegovina, Yugoslavia. Products: Dairy, meat and coarse wool. Pramenka type.

Brownheaded Mutton
(Often called Swiss Brownheaded Mutton). Area: Switzerland. Products: Meat and short wool. Produced from Grabs plus Oxford Down 1938.

Bündner Oberland
Area: Graubünde, Switzerland, Products: Coarse wool. Similar to primitive sheep of Peat type.

Campine
Area: North Belgium. Extinct c.1950.

Carinthian
Area: Austria. Products: Coarse wool, meat and dairy. Lop-eared Alpine type.

Chios
Area: Chios, Greece and Izmir, Turkey. Products: Dairy, meat and coarse to medium wool. Origin unknown.

Cyprus Fat-Tailed
Products: Dairy, coarse wool and meat. Near East fat-tailed type.

Dala
Area: Voss and Hordaland, Norway. Products: Semi-fine wool. Similar to Cheviot and originated 1860-1920 by combining Cheviot, Leicester and Old Norwegian.

Dalmatian-Karst
Area: Yugoslavia. Products: Meat, coarse wool and dairy. Small Pramenka type originated from Dubrovnik with some Merino blood.

Dub
Area: Yugoslavia. Products: Dairy, meat and coarse wool. Bosnian Mountain type.

Dubrovnik
Area: Dalmatia, Yugoslavia. Products: Medium wool and dairy. From Merino × Pramenka in late eighteenth and early nineteenth century.

East Friesian
(East Friesian Milch and the Milch sheep). Area: Germany. Product: Dairy. Now of considerable importance for breeding purposes.

Entre-Sambre-et-Meuse
Area: Belgium. Products: Meat and coarse wool. Extinct about 1950.

Evdilon
Area: Ikaria, Greece. Products: Dairy (milk the principal product). Stall fed, probably a native thin-tailed × Anatolian fat-tailed.

Faeroes
Area: Faeroes. Products: Wool and meat. Icelandic × Old Norwegian.

Finnish Landrace
Area: Finland. Products: Wool and meat. Northern short-tailed type.

Friesian
Area: Netherlands. Products: Dairy. Marsh type similar to East Friesian.

German Blackheaded Mutton
Area: North and West Germany. Products: Meat and short wool. Derived from Hampshire Down + Oxford Down (+ Shropshire and Suffolk) 1870-1914. Also found in Russia.

German Heath
Area: Luneberg Heath, Hanover, North West Germany. Product: Coarse wool.

German Improved Land
(Or Württemberg Merino). Area: South Germany and Alsace Lorraine. Products: Medium wool and meat. From Merino imported late nineteenth century and crossed with Württemberg Land.

German Mutton Merino
Area: Germany. Products: Meat and fine wool. c.1904 from Précoce imported c.1870.

German Whiteheaded Mutton
Area: North West Germany. Products: Meat and wool. From Cotswold × Marsh in nineteenth century. (Also known as Oldenburg.)

Greek Mountain Zackel
Area: Vlakhiko, Greece. Products: Dairy, meat and coarse wool.

Icelandic breed.

Greek Zackel
Area: Greece. Products: Dairy, meat and coarse wool. Several varieties.

Heath (Type)
Area: North West Europe. Product: Coarse wool. Many specific breeds.

Icelandic
Area: Iceland. Products: Meat, dairy and wool. Originally derived from Old Norwegian. Meat accounts for 85% of value of products.

Island Pramenka
Area: Yugoslavia. Products: Coarse/medium wool, meat and dairy. Pramenka type, many variations on the islands of Adriatic coast.

Istrian Milk
Area: East Istria and Croatian Coast, Yugoslavia. Products: Dairy, meat and coarse wool. Pramenka type.

Karagouniko
Area: Thessaly, Greece. Products: Meat, dairy and coarse wool. Greek Zackel type.

Karakachan
Area: Macedonia. Products: Dairy, meat and coarse wool. Zackel type.

Katafigion
Area: North Eastern Greece. Products: Medium/coarse wool, dairy and meat. Small migratory breed of Ruda type.

Kimi
Area: Euboea, Greece. Products: Dairy, meat and medium wool. Local breed derived from Skopelos.

Kosovo
Area: South Siberia. Products: Meat and coarse wool. Pramenka type.

Krivovir
Area: East Serbia. Products: Meat, coarse wool and dairy. Pramenka type.

Kupres
Area: Yugoslavia. Variety of Bosnian Mountain sheep.

Leine
Area: South Hanover, Germany. Products: Wool and meat. Improved by Leicester, Cotswold and Berrichon in nineteenth century.

Lika
Area: Croatia, Yugoslavia. Products: Dairy, meat and coarse wool. Pramenka type.

Lipe
Area: Morava Valley, North Serbia. Products: Dairy, meat and coarse wool. Pramenka type.

Lop-eared Alpine (Type)
Area: Alps. Products: Meat, coarse/medium wool also dairy. Most typical breed Bergamo but many other varieties.

Maltese
Area: Malta. Products: Dairy, meat and coarse wool. East Friesian has been in-

troduced to improve milk yield of this typical three-purpose sheep.

Marsh (Type)
Area: North West Europe. Many breeds in the area.

Mytilene
Area: West Lésvos, Greece. Products: Dairy, meat and coarse wool. Originated from Chios × Turkish breeds similar to Kamakuyruk.

Northern Short-tailed (type)
Area: North Europe. Similar to Heath. Many important breeds of what are generally known as Scandinavian type such as Old Norwegian, Manx Loghtan, Romanov, etc.

Old Norwegian
Area: Islands and Setesdal, Norway. Product: Wool. Origin of Icelandic, Faeroes and improved Norwegian. Now rare.

Oldenberg See German Whiteheaded Merino.

Ovče Polje
Area: East Macedonia, Yugoslavia. Products: Meat, coarse wool and dairy. Pramenka type.

Pag Island
Area: Croatia, Yugoslavia. Products: Medium wool and dairy. From Merino × Pramenka in early nineteenth century, similar to Dubrovnik.

Pälsfår
Area: Sweden. Products: Meat and wool. New sheep recently developed.

Pirot
Area: South East Serbia, Yugoslavia. Products: Meat and coarse wool. Pramenka type.

Pälsfår breed, Sweden.

Piva
Area: North Montenegro, Yugoslavia. Products: Meat, coarse wool and dairy. Pramenka type.

Pramenka (Type)
Area: Yugoslavia. Products: Meat, coarse wool, also dairy. There are many varieties of this important type, of which a full list is given in Mason: *Sheep Breeds of the Mediterranean.*

Privor
Area: Yugoslavia. Products: Dairy, meat and coarse wool. Variety of Bosnian Mountain.

Rhodes
Area: Dodecanese Islands, Greece. Products: Dairy, meat and coarse wool. Probably from Daglic × thin-tailed dairy wool sheep.

Rhön
Area: Central Germany. Products: Wool and meat. Recognised since 1844.

Roumloukion
Area: Central Macedonia, Greece. Products: Meat, dairy and coarse wool. Ruda type.

Ruda (Type)
Area: Balkans. Products: Coarse wool, meat and dairy. Somewhat similar to Zackel.

Rygja
Area: Rogaland, Norway. Product: Short wool. From Cheviot, 1850 onwards, × Old Norwegian with Leicester or Oxford Down blood. Named in 1824.

Šar Planina
Area: West Macedonia, Yugoslavia. Products: Meat, coarse wool and dairy. Pramenka type.

Serrai
Area: Macedonia. Products: Dairy, meat and coarse wool. Ruda type.

Sfakia
Area: Crete. Products: Dairy, meat and coarse wool. Zackel type.

Sitia
Area: Crete. Products: Dairy, meat and coarse wool. This sheep has probably changed little since Middle Minoan times c.1800 B.C.

Sjenica
Area: South West Serbia. Products: Meat, dairy and coarse wool. Pramenka type.

Skopelos
Area: North Sporades, Greece. Products: Dairy, meat and medium wool. Rare and disappearing breed.

Solčava
Area: North Slovenia, Yugoslavia. Products: Meat, coarse/medium wool, also dairy. Late eighteenth and early nineteenth century from Bergamo and Paduan × local giving Seeland and Sulzbach variations of Carinthian.

Steinschaf
Area: Salzburg, Austria; Val Venosta, Italy; South East Bavaria. Products: Meat and coarse wool. Small primitive sheep similar to Bundner Oberland.

Stogoš
Area: South Banat, Yugoslavia. Products: Meat, coarse wool and dairy. Pramenka type.

Svrljig
Area: East Serbia, Yugoslavia. Products: Meat, dairy and coarse wool. Pramenka type.

Swedish Landrace
Area: Sweden. Products: Wool and meat. Northern short-tailed type, now important for breeding purposes.

Swiss Black-Brown Mountain
Area: West Switzerland. *See* Black-Brown Mountain.

Swiss Brownheaded Mutton
Area: North Switzerland. *See* Brownheaded Mutton.

Swiss White Alpine
Area: West Switzerland. Products: Meat and short wool. From 1936 onward from Swiss White Mutton with 50-75% Ile de France blood.

Swiss White Mountain
Area: South and East Switzerland. Products: Meat and short wool. Lop-eared Alpine type.

Texel
Area: Netherlands. Products: Meat and long wool. Marsh type, originally during present century from Leicester, and Lincoln crossed with local sheep. Now an important breed that is coming into the British breeding scene.

Texel breed, Netherlands.

Tsigai
Area: South East Europe. Products: Medium/coarse wool, dairy and meat. Origin of many breeds of South East Europe.

Tyrol Mountain
Area: Austria. Products: Coarse wool and meat. Lop-eared Alpine type.

Valais Blacknose
Area: Switzerland. Products: Coarse wool and meat.

Vojvodina Merino
See Merino section. Area: North East Yugoslavia.

Wilstermarsch
Area: Holstein, Germany. Products: Dairy, meat and wool. Marsh type.

Württemberg Merino
Area: South Germany, Alsace Lorraine. *See* Merino section.

Zackel
Area: South East Europe. Products: Coarse wool, dairy and meat. Important type.

Zante
Area: Zakynthos, Ionian Sea, Greece. Products: Meat, coarse wool and dairy. Originally from Bergamo.

Zeta Yellow
Area: South Montenegro, Yugoslavia. Products: Dairy, meat and coarse wool. Pramenka type.

America

It is in America, both North and South, that first the Merino then the British breeds have to all practical purposes replaced the older, native sheep of which there were in fact comparatively few.

As far as North America is concerned there has been a great decline in sheep numbers. In USA 56 million in 1942 now down to 14 million, and despite noble efforts by sheep and wool organisations this trend is not likely to be reversed. In the following list a few native American sheep are given but although interesting for special reasons, they are not very important. The sheep have always been firmly based on the Merino and the British breeds.

Central America, with Mexico 5,640,000 the main centre, is comparatively unimportant. But with South America we come to a great sheep area, as the following figures show:

	Million
Argentine	42·000
Bolivia	7·326
Brazil	25·500
Chile	6·000
Columbia	2·000
Ecuador	2·020
Peru	17·200
Uruguay	15·902

They are, almost without exception, Merino or British based. No better example could be given of the place these two groups of sheep have than the simple fact of what has happened in South America.

Unlike the position in other countries, there are a number of books about American sheep. I have found the following particularly useful. C. W. Towne and E. N. Wentworth *Shepherd's Empire* 1945, American Sheep Producers Council Inc. *Sheep and Men, An American Saga*, and the sheep section in H. M. Briggs *Modern Breeds of Livestock*. For both parts of the continent I have utilised personal knowledge. In the

North I have found when there, many people—officials of the sheep and wool boards, individual sheep growers, etc., keen to help. With the South during my years as a buyer and user of wool, I both valued and bought South American Merino wool, also the fine crossbreds both the Puntas and the Falklands. Firms engaged in these two trades, notably the (then) Falkland Island Company and the London Wool Brokers, Jacomb Hoare, who played a major part in the founding of the Puntas (i.e. Patagonian) wool trade, have answered questions.

North American Breeds

The typical American wild sheep, the Bighorn, had no effect on American sheep breeding and the story first begins when the sheep that Columbus and those who followed him brought to Central America, moved into the then Spanish ruled area of the West. Although some Merinos must have come, most of these early Spanish sheep seem to have been the coarse Churro type and little fine wool was grown in America until, following the dispersion of the Merino sheep from Spain, the USA shared in the great rush for these sheep. American growers, supported by government action, imported them and this led to large flocks being bred, but American wool played little part in the world market, being all consumed at home. During these years two distinct Merino strains were developed—the Delaine and the Vermont. Later a number of Rambouillet Merinos were also introduced and their effect is clearly evident in what remains of the Merino in the USA. The history of these breeds has been briefly discussed in the Merino section of this book. Then, when indus-

trialisation, and therefore more people brought an increasing demand for mutton, the United States turned to the classic British breeds, several of which had been, and even now, are more numerous here than in Britain itself. The remaining—what may be called original USA breeds in the sense that they do not occur elsewhere are therefore very few and comparatively unimportant. Today the sheep population of the USA is much reduced and the main strain is the Rambouillet Merino, in most cases crossed with British breeds.

American Tunis
Area USA. Product: Meat. Originally from Barbary imported from Tunis in 1799. The breed is now very rare.

Bell Multi-nippled
Area: Nova Scotia, Canada. Developed by Graham Bell in 1890 and bred till 1922.

Bighorn
Area: Rocky Mountains, USA. One of the four breeds of wild sheep that still remain and the only one that does not appear to have had any effect on domesticated breeds. Found today fairly widely scattered over the Rocky Mountains and is certainly, when grazing in its natural habitat, a fine sight.

Canadian Corriedale
Area: Alberta, Canada. Products: Medium wool and meat. Corriedale × Lincoln × Rambouillet 1919–34.

Columbia
Area: Idaho, USA. Products: Meat and medium wool. From 1912 onward by crossing Lincoln with American Rambouillet. Still found in America and has an active breed society.

American Bighorn.

Columbia Breed, USA.

Montadale

Area: Missouri, USA. Products: Meat and medium wool. Cheviot (40%) × Columbia (60%) 1931.

Multi-nippled

Area: USA. 1923-41 from Bell Multi-nippled × (Southdown × Rambouillet) + Suffolk blood.

Navajo

Area: Arizona, New Mexico, Utah, USA. Product: Coarse wool. Originally from the Churro, said to have been the wool that was used for making the famous Navajo blankets. Does not now exist in its original form.

Panama

Area: Idaho, USA. Products: Meat and medium wool. Bred from 1912 onwards from Rambouillet × Lincoln similar to Columbia.

Rocky Mountain See Bighorn.

Romeldale

Area: California, USA. Products: Meat and medium wool. From the 1915 New Zealand Romney × Rambouillet.

Targhee

Area: Idaho, USA. Products: Meat and medium wool. In 1926 'Select Rambouillet rams were mated with select ewes that had been produced from crossing Corriedale rams on crossbred Lincoln/Rambouillet ewes and by mating Lincoln rams with Rambouillet ewes.' (Briggs).

South American Breeds

As the numbers already quoted show, South America is one of the great sheep producing areas and with approximately 118 million sheep is second only to Australia and the USSR. Nowhere is the argument already stated, that the two important themes in sheep history and therefore in sheep today, are the Merino and the British contribution more completely born out. Together they dominate the scene. Historically it was the Merino, now it is more the British. Uruguay remains the centre for the Merino and M.V. Merinos, as they are called, are a distinctive variety and well known in the world wool markets. Everywhere, however, the British based crosses in the main grown for their meat, are much in the majority. Only a few breeds call for individual mention here, two of them specialised and rare, the other two the Puntas and Falklands although ultimately British based, are so distinctive that they call for separate mention.

Brazilian Woolless

Area: Morada Nova, Ceara, North East Brazil. Products: Meat and dairy. From West African or Bardalero Churro.

Crioulo

Area: Brazil. Products: Meat and coarse wool. From Bordalero.

Falkland

Area: Falkland Islands. Product: Wool

(fine crossbred 58s (24 microns). Developed over the years from Cheviot but with much crossing with Lincoln and Romney, as well as Merino, which accounts for fineness of wool. A very distinctive sheep.

Puntas
Area: Patagonia, Chile. Products: Wool and meat. Similar to Falkland from whence the sheep originally came. For a full account of this interesting breed see Ponting *The Wool Trade*.

5 SHEEP PRODUCTS

The dual, even multi-purpose of the sheep both past and present has been sufficiently stressed in the history and present day strains of the individual breeds. Here it only remains to state the continuing position, as far as one dare forecast. Wool and meat will remain the two main reasons why sheep continue to be produced. The small demand for the milk products, almost entirely for such cheeses as Roquefort, will provide a small market and in special cases, there is the value of the skins as for example with the Karakul sheep. The pure bred sheep for export, of course, also remains very important.

Wool and Allied Products

During the past thirty years wool has lost ground to the synthetic fibres in the sense that the total percentage of wool used as textiles (both for wear and for furnishing) has declined but not, as the figures show in the appendix, the *actual* weight which has only fractionally declined. The quantity of synthetics, both rayon and the more completely synthetic like Nylon and Terylene have already greatly increased.

During the early nineteenth century there was certainly a basic shortage of good quality wool, partly solved by the development in Australia, New Zealand and South America of wool growing, and partly by the ever increasing use of torn up wool rags (shoddy and mungo). Great skill was shown by the manufacturing trade in processing and making serviceable cloths from these materials. The trade in relatively low priced woollen cloths has been largely replaced by the new synthetics—particularly in Britain, where the low woollen trade of Batley, Dewsbury and Morley has largely disappeared. But the demand for the new wool fabrics continues and the middle decades of the twentieth century have seen an increasing demand for all natural products. Wool is certainly one of the most natural of all.

The wool products of the sheep can be considered under the following subdivisions:

Where wool is the *main* product, subdivided into

    (a) fine wool—i.e. Merino
    (b) coarse wool, mainly the so-called East Indian wool for carpets

Where wool is a *by-product* from meat production, and here there are two very distinctive sub-groups:

    (a) where it is shorn from the living sheep
    (b) where it is removed from the dead skins of the lamb or sheep

Main Product Wool

(a) Fine

When discussing the historical development of the sheep and that of the individual breeds, the importance of the Merino sheep for giving fine wool has been stressed. It remains the one source

Wool Bales after inspection by buyers.

and here it is only necessary to sketch its main manufacturing uses. To a large extent these are entirely apparel—the many non-apparel uses for wool, carpets, upholstery, miscellaneous coverings, etc., are almost all supplied from the non-Merino, i.e. Crossbred or so-called carpet sections. Merino wool is apparel wool and in the past has been converted into fabric by weaving, and knitting, and this has increasingly been done by the so-called worsted method of manufacturing, not by the woollen.

Few words have so tended to confuse the outsider as woollen and worsted. Both use wool but, as stated, today the worsted section uses most of the best wool. In the past this was not the case; until the middle of the nineteenth century all fine wool, that is Merino, went

into woollen manufacturing. It was only when satisfactory mechanical combing (i.e. worsted) machinery was invented and rather longer Merino wool began to come from Australia that the worsted section of the industry moved into its present position.

The difference between woollen and worsted today resides essentially in the choice of wool and the method of yarn making. In general terms, long wools are used for worsted, short for woollens, but length with wool is relative—8 cm (3 in) is long for Merino, short for Crossbreds, but short wool under 5 cm (2 in) is not really suitable for combing. During the yarn making the short wool that is removed with the long on any sheep is, in worsted yarn manufacture, removed by a process of combing. The resulting short fibre, the noils as they are called, are incidentally used by the woollen trade. During this combing process and during the following drawing process, every effort is made to produce a yarn in which the fibres will lie parallel, yielding a strong, non-hairy yarn and resultant fabric. With woollen yarn it is quite different. The wool is short anyway, and every effort is made to retain these short fibres in the yarn and no parallelisation is sought. There is no combing, only carding, and the result is a soft hairy, kindly, warm feeling fabric but one that will lack the strength of the worsted. In the past woollen cloths were heavily fulled, that is shrunk, and worsteds were not, but this one very distinctive difference hardly applies today.

The worsted yarns so produced from Merino can, as stated, be converted into fabric by either weaving or knitting. During the second half of the twentieth century and indeed, until the end of the

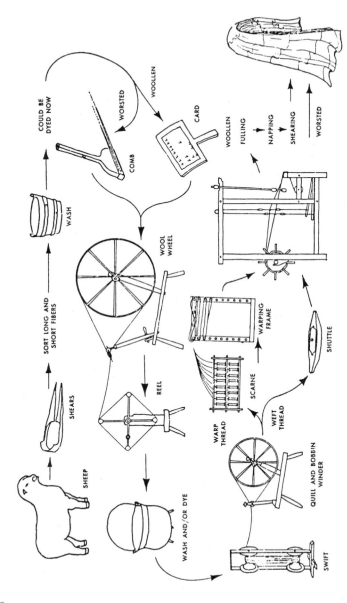

Process chart showing how wool is converted into fabric. (From *Textile History* Vol III 1976, reproduced from M. L. Channing *The Textile Tools of Colonial Homes*.)

second half of the Second World War, a considerable amount (in Britain, at least one third) of worsted yarn went into the knitting trade. This trade is now rather reduced, somewhat oddly because in the past thirty years knitting has overtaken weaving as the main method of producing pile fabrics but unfortunately for wool producers, the modern knitting machines call for very strong yarns and these can only be obtained from the synthetic fibres.

On the other hand, the position of Merino wool in the woven apparel trade appears secure.

(b) Coarse

Although most non-Merino wool comes from sheep that are bred for meat purposes, this is not true in every case. In Asiatic areas considerable numbers of sheep produce coarse wool with the carpet trade the main object. These wools are often described in the wool trade as East Indian—a rather misleading phrase. They come from Asia as well as India. Carpets are manufactured so that a pile shows on the surface and good wearing depends on the resilience of the individual wool fibres to spring back. Wools with a mixture of really coarse fibre and some fine (but not very fine) wool give the best result. For quite a time after synthetic fibres had made new advances into apparel cloths, they were comparatively unsuccessful with carpets, but with the coming of the acrylic fibres this has rather changed; however, coarse carpet wools are still widely used, often now mixed with synthetics. Although the demand for carpet wool which led many New Zealand growers to try to

Members of the Paiyinwula Production Brigade, Inner Mongolia, shearing.

Handshearing in Wales.

develop distinctive qualities in their Romney wool may not be quite as buoyant as it was, these speciality-produced carpet wools are certain to be needed in the future. Although still a major world trade, many carpet wools are being processed where they are grown in Central Asia which was, of course, the original home of the carpet.

By-Product Wool

(a)   Shorn from Living Sheep

To turn now to the large amount of wool—well over half of total supply—which is produced from sheep basically bred for meat. All British sheep breeders have this meat production in mind and the importance of British sheep in the world rests on the fact that they are the basis of this important section of the trade.

These wools are often called crossbred wools—sometimes Cheviot wool—both names are quite misleading and call for some explanation. Many of the sheep producing these wools are themselves crossbred. This is well illustrated by the British position today, where most of the sheep we see are crossbred. Indeed, a fascinating, even complicated, system of cross breeding has been established to get the ideal fat lamb. To take a typical example, the Scottish Half-bred (Cheviot crossed with Border Leicester) clearly produces crossbred wool but to the woolman the wool from components of that breed, the Cheviot and the Border Leicester, also produces wool that he would describe as crossbred in that they are not Merino. It should be made clear that as far as the woolman is concerned the wools from the Half-bred and the Border Leicester are not all that

different and there is no reason why he should be concerned with the breed. The complicated breeding pattern established in Britain is not so widely adopted elsewhere. In New Zealand for example, the Romney in a straight, non-crossbred form, holds premier place and to emphasise the point made above, New Zealand, with its large quantities of Romney wool, is regarded as the main producer of crossbred wool.

(b)   Dead Skins

This great quantity of crossbred wool can be further sub-divided into wool which is cut from the living sheep, and wool which is taken from the lamb or sheep that has been killed for meat. The first group, usually called shorn wool, is produced from sheep that have been kept to obtain the lambs for killing or from lambs that are shorn before they are slaughtered. It is an important section of the trade in this type of wool. But in my opinion it is in the other section, namely the wool taken from the skin of sheep killed for meat, that gives a distinctive 'feel' to this trade where indeed New Zealand slipes have always had a notable place. As most sheep meat is now lamb meat, the wool is mostly lambs' wool and arrives in a wide range of lengths according to when the animal is killed.

There are two distinct methods of removing the wool from the skin of the sheep:

Chemical method   the back of the skin is painted with a lime and sulphide mixture. The compound works through the skin, weakens the root and allows the wool to be pulled off. This lime method leaves the skin in a good condition for tanning but may result in damage to the wool.

Bacterial method advantage is taken of the bacteria already present in the skins which are wetted and hung up in a warm room for several days. Under these circumstances the bacteria multiply and rot away the roots of the wool fibres until they can be easily pulled away from the skin. The result is an indifferent skin for tanning but a superior wool.

Meat (Lamb and Mutton)

The production of meat is the main reason for sheep farming today. About 60% of sheep are reared primarily for meat. Britain, New Zealand, and much of South America, three of the five great sheep producing areas, are in this position compared with Australia and South Africa where wool takes preference. Although many sheep—European, Indian and Asiatic—are largely bred for meat, in most cases they are for home consumption, and the great world lamb and mutton industry is to a large extent based on the British breeds, and it is necessary to emphasise yet again that it is this that gives these British breeds their great importance among sheep of the world today.

Another point worth mentioning is that whereas the essential meat breeds produce useful wool, the main wool producing breed, the Merino, yields a rather indifferent meat. The toughest mutton I have ever eaten was from a Merino ewe in Australia, the best perhaps from a young hogget (Romney) in New Zealand.

As far as meat is concerned, the products of the sheep can be divided into lamb and mutton. The increasing precedence of the fat lamb trade is the development of the second half of the nineteenth century (especially from New Zealand). Previously, lambs were too precious to be killed off as they are now.

To say a little about the distinctive qualities of lamb and mutton. A sheep is a lamb until it has acquired its first pair of permanent teeth, when it becomes a hogget. From the point of view of culinary usage, a further division can be made into either a baby lamb, or a lamb proper. The first is milk fed, in other words fattened off the mother without weaning. The flesh is white and tender, rather lacking in taste but a real, and now expensive, gastronomic delight, as all who have, for example, eaten baby lamb in Greece, will agree. In many countries including Great Britain, the term lamb covers both lamb as defined above and hoggets, in other words any sheep not yet twelve months old. It then becomes mutton. The great gastronome André Simon, wrote some years ago: 'In olden times however, mutton was the meat of wether or ewe when three, four or five years old; it was darker in colour, better in flavour, with plenty of fat, a very much finer meat than that from the old lambs or yearling sheep of the present day. In the USA mutton is not mentioned at all in polite culinary society, one never gets anything but lamb.' *A Concise Encyclopedia of Gastronomy* Section VIII, Meat (1947). Today it would also have probably been frozen and these two facts have partly accounted also for the comparative unpopularity of sheep meat in any form in the USA.

With mutton different breeds do give different qualities of meat but this hardly

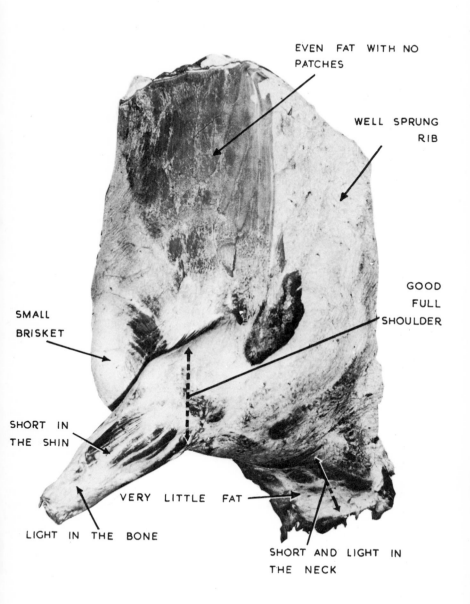

EVEN FAT WITH NO
PATCHES

WELL SPRUNG
RIB

GOOD
FULL
SHOULDER

SMALL
BRISKET

SHORT IN
THE SHIN

VERY LITTLE FAT

LIGHT IN THE BONE

SHORT AND LIGHT IN
THE NECK

Features of the ideal lamb carcase.

applies to the lamb, which will almost certainly, if bought in a shop, be of a crossbred nature and to indicate what should be looked for a government report quoted by Simon may be given: 'In a lamb or young mutton carcase of high quality, the bone should be fairly soft, the flesh light in colour, fine in grain, firm, well developed and full of sap. The fat of a certain colour and well distributed and the conformation good. The relation of the muscle to gristle and bone should be relatively high.' It should not have been frozen. The best joints— and this applies equally to mutton—are the saddle and the shoulder but they are the least economical, carrying more bone to meat than the loin, the legs and the neck. Chops, of course, come from the loins and there are also the various incidentals, the odds and ends from the tail to the head. This type of sheep meat is that supplied today. Turning to mutton, the best tasting would come from a well fed sheep of three or four years old, but for the best result it should be hung for two or three weeks after slaughter but not frozen. Breed then becomes important and the position was well put by the poet, Peacock, in the lines already quoted. These lines, if changed to 'the valley sheep are fatter, but the mutton sheep are sweeter, we therefore deemed it meeter to dine off the latter' would be sound gastronomical advice, especially if in mutton one includes the hill and down breeds. The best and most distinctive mutton comes from the mountain or hill breeds like the Blackface, and the Down, above all the Southdown. The meat from the so-called long woolled breed, the Leicester, is not so good. In this respect Bakewell has something to answer for.

Outside Britain and to some extent, Northern Europe, lamb and mutton have never had the popularity of beef. This is noticeable in America where a steak is regarded as the ultimate, even more so than in England. This view is unfortunate, both for the farmer (the producers) and for the eater (the consumer). Simon in the volume already quoted, gives as many classic recipes for lamb and mutton as for veal and beef, actually between 90 and 100 of each.

Milk (and Cheese)

In the past, for example in England in Saxon times, sheep were reared mainly for their milk producing qualities but this is no longer the case and although sheep's milk is still drunk it is hardly a product of commercial importance, and in this small but interesting section on the products of the sheep, it is sheep's milk as a raw material for cheese that most calls for comment. In the lists of breeds the group that are kept in South West France to produce Roquefort cheese has been noted. But Roquefort, although the best known, is not the only good cheese produced from sheep's milk. Its fame however means that it should be mentioned first. Roquefort can only be made during the lambing season, that is for at most, five months of each year. The name comes from the village of Roquefort in the delightful Rouvergue country of the Cevennes in the remote Aveyron Departement. It can also only be made there partly because of the suitable sheep grazing land, but also because the caves of Roquefort play a leading part in maturing the cheese. The cheese 'is made from the evening and morning milk mixed, and the rennet which is taken from lambs'

Cheese market, Alvani province, USSR.

stomachs instead of calves, is added when the temperature of the milk has been raised to 90°F. The curd is coagulated very quickly and it is piled up in layers, with some mouldy breadcrumbs between each layer. It is then pressed and matured in natural caves or caverns where the temperature is about 46°F and the humidity very great, owing to the existence of an underground lake under the caves. The mouldy breadcrumbs are a culture of the same mould used for the making of Stilton, the *penicillium glaucum*, which grows more rapidly, being already in full activity when introduced into the curd.' (See Simon *Encyclopaedia of Gastronomy* Section IX, Cheese.)

Other sheep milk cheeses which the traveller finds locally include the Italian Pecorino, the Portuguese Serra de Estrella, the Sicilian Canelotti and the Italian Riccotta Romana, and the Greek Feta cheese.

Miscellaneous, especially skin products

These are essentially uses made of the sheepskin, the wool that is taken from the skin has already been described but the wool can, of course, be left on the skin to give the well known sheepskin coat or sheepskin rug. This leads to the most distinctive, if not the most important, use of this kind and one which has led to the breeding of a special kind of sheep, especially in Russia—the Karakul, also found in South Africa. Here the object of the production is the lambskin. With the Karakul it is the new born lamb that is sought. The pelt is taken from the lamb which is killed within twenty-four hours of its birth. The re-

Karakul sheep skins, USSR.

sulting product, known commercially as the Persian Lamb, or Astrakhan (although this word sometimes describes a woven imitation) is an expensive and luxury product. Considerable developments in the breeding of these sheep have taken place in Namibia (the old South West Africa) and in the USSR where increased lambing rates—obviously of key importance for a commercial product such as this—have been obtained. The USSR has also developed another breed, the Romonov, but here the lambs are not killed as young. The high lambing rates of the Romonov may well give this new breed other significance. Then of course there is the skin of the sheep after the wool has been removed, and it should be remembered that the skin is often of less value than the wool. The sheepskin is comparatively poor after the wool has been removed and open textured as compared with, say, the skin of the ox and cow. The younger the skin the better, as the popular kid gloves often made from sheepskin indicate. The skin for these must be milk lambs as, once the lambs have taken to grass, a harshness comes about.

Folding and Breeding

There is one historical use of sheep which, though it has lost a little of its importance, yet remains, namely manure. Before artificial fertilisers were introduced this was vital and as indicated under the individual breeds, the Old Wiltshire Horn sheep was bred almost exclusively for this purpose and the corn economy of the chalk hills of south England depended for its existence on the sheep. As a Yorkshire sheep farmer once said, 'Sheep is gran' things. They

ligs (= lay or spread) their awn moock an troddle (= tread) it in t'ground'. (quoted by E. Sandars in *A Beast Book for the Pocket* (1937). Although the widespread use of artificial fertiliser has restricted this use with many small flocks, its value still remains.

One final and outstanding product of sheep farming remains and it brings us back to a central theme of this study —the importance of the pure bred. I refer to the rearing of pure bred sheep, mainly rams, for stud purposes. The vital importance of the spread of the Merino and British breeds has already been stressed and it yet remains. The existence of the fine wool sheep in Australia rests on a relatively few great stud farms to which other growers can go to replenish the strain they are wishing to maintain. With British breeds the demand for them remains strong, indeed with some breeds, e.g. the Hampshire, it is their most important use.

6   SHEEP FARMING TODAY

Sheep breeding has changed less than most forms of agricultural husbandry. The sheep have certainly not disappeared from the countryside as has the fowl. Improved genetical studies have brought a certain amount of artificial insemination which may have gone further in some countries such as the USSR than others. Certainly in the great sheep breeding areas of Australia, New Zealand and Britain traditional methods survive, and the shepherd's year remains much as it has always been. Obviously there have been many improvements, in particular lambing rates are much im-

A spray is put on foster ewe and orphan lamb giving them the same smell.

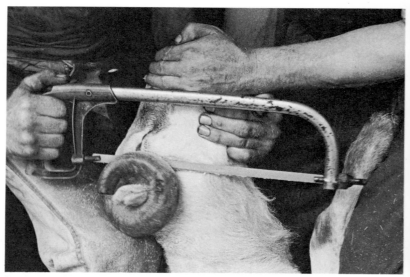

Cutting horn to prevent ingrowing.

roved and one no longer follows the dvice of Gervais Markham who, in his *Cheap and Good Husbandry* of 1614, wrote 'If you would have your Ewes ›ring forth Male Lambs, note when the North wind bloweth, and driving your Flock against the wind: let your Rams ·ide as they go, and this will make the Ewes conceive Male Lambs: so likewise, f you would have female Lambs put your Rams to the Ewes when the wind ›loweth out of the South'.

Indeed, when one reads Elizabethan authors on animal breeding, one cannot but be struck by the fact that their knowledge seems to have been no more than that of the Romans fifteen hundred years before. No Elizabethan writer on sheep gave the farmer as good advice as the great Roman poet Virgil, who when emphasising the Roman desire to produce white wool, wrote:

Is wool thy care? Let not thy cattle go
Where bushes are where burrs and
    thistles grow,
Nor into ranker pasture let them feed;
Then of the purest white select thy
    breed.
E'en though a snowy ram thou shalt
    behold,
Prefer him not in haste for husband of
    thy folds;
But search his mouth; and if a swarthy
    tongue
Is underneath his humid palate hung,
Reject him, lest he darken all thy flock;
And substitute another for thy stock.

The crucial work of Bakewell and Ellman has been sufficiently emphasised and modern sheep farming methods remain to a large extent based on good practice, helped rather than formed by scientific technical knowledge.

To look in a little more detail at the future of the two main products—meat and wool—the latter has perhaps been rather more at risk. The threat from the synthetic fibres has been considerable but from the wool standpoint may well be over the worst. Wool has survived. No synthetic fibre has the same properties as wool. Synthetic fibres, the manufacturers say, are better; what they really mean is that they are different. But what are the advantages of wool? There are many, and amongst those I would place first are the great variety of wool produced which enables manufacturers to make so many different fabrics. It would indeed be a very dull world if all we had were the synthetics. Then there are the natural properties of the fibre, two are outstanding, namely crimp and handle. Makers of man-made fibres have gone to very great trouble to put into their machine-made products the crimp that is natural to wool. Handle, which to some extent depends on crimp, is even more important and I have always regarded the handle of wool as one of its greatest advantages. Phrases that have become common to our language, such as *as soft as wool*, prove that to most people this is indeed the case.

When one turns to processing, the outstanding advantage is the fact that one has a fibre which naturally contains a considerable quantity of water and that as a result there is little trouble from static electricity. People who have been in the business of processing wool and turn to synthetics, quickly learn the difference. There is also the important fact that wool does not burn easily and so therefore, there is less risk of fire. It has, in fact, been surprising that wool growers have not made more of this point.

Another property not mentioned as much as it should be is the ease of dyeing wool compared with man-made fibres. This at least partly compensates for the comparatively high cost of the fibre. Finally, in the field of processing, the wide range of finishes that can be given to wool are not sufficiently praised. To the user the capacity of wool to absorb moisture makes for most of the comfort. Such are the advantages of wool, and most of the disadvantages can be overcome if due attention is given to proper processing. Indeed, to some extent use can be made of them. Perhaps the greatest disadvantage today is the fact that wool shrinks but one should not forget that it is because wool shrinks that felting is possible, and that by using this felting property, warm fabrics can be made.

The various types of vegetable faults that are found in wool without doubt constitute one of the major faults of the fibre. This is much more the case now than in the past. The arrival of synthetic fibres which are supplied free from impurity has given the trade new standards and increasingly fibre users base their ideals on the standards so established.

Meat has been less at risk, there has been no successful synthetic meat and for the foreseeable future lambs will continue to be in demand, not quite in the same way as beef-steak but sufficiently to keep the quantity of meat producing sheep at least at their present level.

Consequently the developing but basically underlying pattern of sheep farming briefly outlined here will probably continue for at least the life of those alive today.

Finally, what are the possibilities of increased production of wool and meat? There are several ways of attempting this which can be divided into two main groups:

(a) An increased number of sheep, which can be achieved by
    (i) heavier stocking
    (ii) the use of new country
(b) It is also possible to increase the yield of wool and meat per sheep, which can be achieved by:
    (i) breeding bigger sheep
    (ii) giving more attention to culling for weight of fleece and carcase
    (iii) using scientific knowledge of genetics to increase lambing rates

As far as Western Europe and the USA are concerned, an increased number of sheep seems unlikely. Australia and New Zealand may be expected to reduce their figures slightly. The same probably applies to South America and without doubt the USSR and China are the most likely places for increases. Even if these two countries successfully increase industrial production there is still plenty of land for more sheep. Overall it seems that the larger number of sheep in the two latter areas will about balance the decline elsewhere.

Expansion, therefore, must come from the yield per sheep and here there are some good opportunities. In so far as fine wool sheep are small sheep, the comparatively declining demand for fine wool will inevitably mean—as is already occurring in Australia and South Africa—a switch to the broader woolled, heavier Merinos, and probably to a wider use of the British based breeds. The present proportion of wool to meat breeds of approximately 40/60% will probably change to, say $33\frac{1}{3}/66\frac{2}{3}$%. The use of better culling, which to a large extent depends on better records properly analysed, will clearly help and the interest in, and work done by, using lambing rates in genetical studies in all their various forms must surely mean a considerable increase, maybe one of 10/15% in wool and meat yields in the last decades of the twentieth century.

One final point may be made, bigger sheep could of course be obtained by breeding sheep on better lands but the whole history of sheep production means that if one moves one's sheep to better land one then comes into conflict with cattle raising, which is usually a more profitable farming occupation, always assuming that the land is suitable. Sheep have essentially been the ideal animal for relatively poor land and as poor land decreases, so sheep find it more difficult to survive economically speaking.

Shepherds and their Dogs

Few human and animal themes have so delighted the European poet as the shepherd and his shepherdess, the sheep themselves and the sheepdogs. It was a major theme in Greek literature and even more so in our own Elizabethan. Some useful information can be obtained from these sources; as far as English literature is concerned Drayton's *Polyobilon* has already been quoted with reference to the Ryeland or, as it was then called, the Leominster. Those interested in the byeways of English literature might find it worth while to look at a minor eighteenth century poem called 'The Fleece' by John Dyer. Somewhat unusually, the technical details here are excellent, the verse

Sheepdogs guarding flocks, USSR.

perhaps less so. It is incidentally an important source for information about some of the early machine inventions. But above all it was the Elizabethans who best sang of the shepherds and their friends, both human and animal. The romantic stories they told had little connection with reality. For reality, W. H. Hudson's wholly admirable prose work *A Shepherds Life*, should be read or alternatively, the minor Ettrick poet, James Hogg's prose work on sheep. However, with writing in the literary genre the shearing scene in Shakespeare's *The Winter's Tale* is the place where the pastoral shepherd convention reaches its greatest height. More realistically, it is worth considering how and why this theme of the shepherd and his sheep have affected so much more than say the cowman and his cow. Perhaps the sheep dog has been mainly responsible. The dog was the first animal to be domesticated, his association with man goes back to pre-Neolithic hunting days and represents a unique partnership. Gosset, in *Shepherds of England*, has brought together much information on this theme.

BIBLIOGRAPHY

British Agricultural History Society, Canterbury (1952–1978) *Agriculture History Review* Vol. I–XXVI

British Wool Marketing Board, Bradford (Revised edition 1978) *British Sheep Breeds—Their Wool and its Uses*

Carter, H. B. (1964) *His Majesty's Spanish Flock: Sir John Banks and the Merinos of George III of England* Angus and Robertson, London

Coleman, J. (1887) *Cattle, Sheep etc., of Britain* Horace Cox, London

Epstein, H. (1969) *Domestic Animals of China* Commonwealth Agricultural Bureau, Farnham Royal

Gosset, A. L. J. (1911) *Shepherds of Britain* Constable, London

Klein, J. (1920) *The Mesta 1273–1836* Cambridge (Mass), USA

Low, D. (1845) *On the Domesticated Animals of the British Isles* (no publisher stated), Edinburgh

Luccock, J. (1805) *The Nature and Properties of Wool* (no publisher stated), London

Markham, G. (1614) *Cheape and Good Husbandrie, etc.* (no publisher stated), London

Mason, I. L. (2nd edn. 1969) *A World Dictionary of Types and Varieties of Livestock Breeds* Commonwealth Agricultural Bureaux, Farnham Royal

Mason, I. L. (1967) *Sheep Breeds of the Mediterranean* Food and Agricultural Organisation of the United Nations Commonwealth Agricultural Bureaux, Farnham Royal

National Sheep Breeders Association (Revised edition 1976) *British Breeds*

Ponting, K. G. (1961) *The Wool Trade Past and Present* Columbine Press, Manchester

Power, E. (1941) *The Wool Trade in English Medieval History* Oxford University Press

Shann, E. (1930) *An Economic History of Australia* Cambridge University Press

Trow Smith, R. (1957) *A History of British Livestock to 1700* Vol. I Routledge & Kegan Paul, London

Trow Smith, R. (1959) *A History of British Livestock, 1700–1900* Vol. II Routledge & Kegan Paul, London

Wrightson, J. (1908) *Sheep Breeds and Management* Vinton & Co., London

Youatt, W. (1837) *Sheep: their breeds, management, and diseases, etc.* Society for the diffusion of Useful Knowledge, London (Library of Useful Knowledge)

Zeuner, F. E. (1963) *A History of Domesticated Animals* Hutchinson, London

Ziegler, O. L. (ed.) *The Australian Merino* New South Wales Sheepbreeders' Association

APPENDIX

World Sheep Population (million)

	1960/ 61	1970/ 71	1972/ 73	1973/ 74	1974/ 75	1975/ 76	1976/ 77p
Argentina	46·0	39·0	40·0	39·0	38·0	37·5	37·0
Australia	152·7	177·8	140·1	145·2	151·7	148·6	135·4
Brazil	18·2	24·3	25·0	25·5	26·0	25·0	25·1
Bulgaria	9·3	9·7	9·9	9·8	9·8	10·0	9·7
Canada	1·5	0·9	0·8	0·8	0·7	0·6	0·5
Chile	6·3	5·8	5·4	5·1	5·9	6·0	6·1
China	59·0	59·0	59·0	59·0	59·0	59·0	59·0
France	11·5	10·2	10·2	10·3	10·6	10·7	10·9
Greece	9·4	7·5	7·9	8·2	8·4	8·4	8·5
India	40·2	40·7	40·4	40·2	40·0	40·0	40·0
Iran	28·0	34·0	33·8	34·0	34·0	35·0	35·3
Iraq	8·5	13·1	14·0	15·5	15·5	15·8	15·5
Irish Republic	4·5	4·2	4·3	4·1	3·8	3·5	3·5
Italy	8·2	7·9	7·8	7·8	8·0	8·2	8·2
Lesotho	1·5	1·7	1·7	1·6	1·6	1·6	1·6
Mongolia	13·8	13·3	13·7	14·1	14·5	14·9	14·6
Morocco	15·0	11·1	13·2	12·2	13·0	12·4	12·7
New Zealand	48·5	58·9	56·7	55·9	55·3	56·4	58·2
Pakistan	10·3	10·3	17·5	18·1	18·1	18·1	19·5
Peru	16·0	17·1	17·2	17·1	17·3	17·0	17·3
Romania	11·5	13·8	14·5	14·3	13·9	13·9	14·8
South Africa	33·8	28·9	29·6	28·1	28·7	29·6	30·5
Soviet Union	133·0	138·1	139·1	142·6	145·3	141·4	139·8
Spain	22·6	18·4	17·2	16·3	16·3	15·7	15·6
Turkey	34·5	36·5	38·8	40·1	40·5	41·4	41·5
United Kingdom	29·0	26·0	27·9	28·5	28·3	28·2	28·1
United States	33·0	19·7	17·7	16·4	14·5	13·4	12·8
Uruguay	21·5	18·5	15·9	15·4	14·9	16·0	16·2
Yugoslavia	10·9	8·7	7·8	7·9	8·2	7·8	7·5
Other	86·8	94·7	90·3	93·0	94·2	96·4	98·1
TOTAL	925·0	949·8	915·3	926·1	936·0	932·5	923·5

Sources: IWTO, Commonwealth Secretariat, International Wool Study Group, IWS

Meat (Mutton-Lamb) Production (in metric tons)

	1961–5	1976		1961–5	1976
World	348,570	364,380	Europe	64,526	70,132
Africa	39,196	47,383	Australia	33,305	32,694
North America	17,820	8,628	New Zealand	29,054	33,634
South America	19,895	20,054	USSR	71,645	60,000*
Asia	73,127	91,853			

*Estimated.
Sources: The Meat and Livestock Commission, Milton Keynes.

World Wool Production—Greasy (million kg)

	1960/61	1970/71	1972/73	1973/74	1974/75	1975/76	1976/77	1977/78
Argentina	195	200	177	180	184	188	182	180
Australia	737	891	735	701	794	754	703	680
Brazil	23	32	33	33	34	35	35	35
Bulgaria	21	29	31	32	33	34	33	32
Canada	4	2	2	2	2	2	1	1
Chile	22	22	18	18	17	17	17	17
France	26	20	21	21	22	21	22	22
Greece	11	8	8	9	9	9	11	11
India	35	33	35	35	35	35	35	35
Iran	42	51	44	48	49	50	50	50
Iraq	13	16	17	16	17	18	18	18
Irish Republic	11	10	10	10	10	10	9	9
Italy	15	12	12	11	11	12	12	12
Lesotho	3	4	4	3	3	3	3	3
Mongolia	19	19	20	20	20	20	20	20
Morocco	15	18	18	23	19	20	21	21
New Zealand	267	334	309	285	294	312	303	312
Pakistan	20	20	20	20	22	29	29	29
Peru	10	13	10	10	10	11	11	11
Portugal	11	14	12	14	14	14	14	14
Romania	22	30	31	31	31	32	32	31
South Africa	144	123	114	113	115	114	111	113
Soviet Union	352	419	420	433	462	467	436	458
Spain	38	34	32	32	31	29	29	29
Turkey	47	47	50	52	52	53	54	54
United Kingdom	55	46	48	49	50	49	48	46
United States	147	85	80	72	65	59	53	51
Uruguay	82	78	60	60	63	68	64	62
Yugoslavia	14	12	10	10	10	11	11	10
Other	166	167	16	165	165	165	166	166
TOTAL	2,567	2,789	2,544	2,508	2,641	2,641	2,533	2,533

Sources: IWTO, IWS

World Wool Production—Clean (million kg)

	1960/61	1970/71	1972/73	1973/74	1974/75	1975/76	1976/77	1977/78
Merino	560	604	529	510	562	545	511	504
Crossbred (mainly apparel)	627	706	644	629	668	671	638	644
Other (mainly carpet type)	293	295	299	301	307	308	311	312
TOTAL	1,480	1,605	1,472	1,440	1,537	1,524	1,460	1,460

Sources: IWTO, IWS

INDEX

Numbers in **bold** refer to the text
descriptions of the main breeds;
numbers in *italic* refer to illustrations.